Dedicated to Sir David Attenborough, who brought
wonder and mystery into my childhood.

Adventures in Cryptozoology

Hunting for Yetis,
Mongolian Deathworms,
and Other
Not-So-Mythical Monsters

VOLUME I

Richard Freeman

CORAL GABLES

mango

Copyright © 2019 Richard Freeman

Published by Mango Publishing Group, a division of Mango Media Inc.

Cover Design: Morgane Leoni
Layout & Design: Morgane Leoni

Mango is an active supporter of authors' rights to free speech and artistic expression in their books. The purpose of copyright is to encourage authors to produce exceptional works that enrich our culture and our open society.

Uploading or distributing photos, scans or any content from this book without prior permission is theft of the author's intellectual property. Please honor the author's work as you would your own. Thank you in advance for respecting our author's rights.

For permission requests, please contact the publisher at:
Mango Publishing Group
2850 S Douglas Road, 2nd Floor
Coral Gables, FL 33134 USA
info@mango.bz

For special orders, quantity sales, course adoptions and corporate sales, please email the publisher at sales@mango.bz. For trade and wholesale sales, please contact Ingram Publisher Services at customer.service@ingramcontent.com or +1.800.509.4887.

Adventures in Cryptozoology: Hunting for Yetis, Mongolian Deathworms, and Other Not-So-Mythical Monsters: Volume I

Library of Congress Cataloging-in-Publication number: 20199356743
ISBN: (print) 978-1-64250-015-8, (ebook) 978-1-64250-016-5
BISAC SCI070000—SCIENCE / Life Sciences / Zoology / General

Printed in the United States of America

Adventures
in
Cryptozoology

Praise for *Adventures in Cryptozoology*

'Richard Freeman must surely be the world's most widely-travelled field cryptozoologist, scouring the globe for well over twenty years in search of such elusive mystery beasts as British lake monsters, Mongolian death worms, Tasmanian wolves, Russian and Sumatran man-beasts, South American water tigers, and African dragons, to name but a few. Now he has drawn upon the extensive knowledge and experience gained during his many expeditions and voracious reading to write this fascinating book, packed with new, original insights and forthright opinions, making it essential reading for everyone who dreams of following in his footsteps seeking unknown animals.'

—Dr Karl Shuker, zoologist and author

'Freeman's book on monsters is a page-turning treat. Not only does he pack it with excellent overviews, facts and details on cryptozoology, he also grips the reader with accounts of his own expeditions. These globe-trotting journals bring the hidden and mysterious creatures of nature into the mud and dirt of serious, real-world research, leaving us with a thrilling and inspiring book. Yes, Freeman takes monsters seriously, but he never loses his sense of fun, wonder and adventure. Recommended.'

—Peter Laws, author of *The Frighteners: Why We Love Monsters, Ghosts, Death, and Gore*

Table of Contents

Foreword

Here be dragons...

Had Richard Freeman been living in medieval times, I am quite certain he would have been burnt at the stake for being a wizard with a mind crammed full of arcane knowledge about the strangest, weirdest, scariest, and in some cases possibly even the most dangerous creatures this world has ever seen. But, as he is in fact living in present day Exeter, he has to make do with being a well-known, and to some extent even notorious (something I suspect he would actually be a little bit proud of), figure in the weird and wonderful world of cryptozoology.

To say Richard is passionate about cryptozoology, would be to engage in an understatement of almost criminal dimension. And if you combine this with an almost encyclopaedic knowledge of everything from thylacines to the many varied forms of big, hairy ape-like creatures roaming the world, and the many creatures in between (dragons being especially close to his heart), you are in for several entertaining hours, should you find yourself seated across from him over a couple of pints, or amongst the audience of one of his lectures.

I first met Richard in the mid-1990s at the first UnConvention held by the *Fortean Times* magazine in London, and got to know him later at a number of the Weird Weekends held by the Centre for Fortean Zoology in Northwest Devon. Despite our differences—me being a traditional university-educated zoologist/scientist, a class of beings for which Richard can only muster a fairly limited amount of respect—we became good friends. A situation which, over the years, has resulted in me receiving mysterious and often evil-smelling letters and parcels whenever Richard has been on one of his travels to far-flung corners of the earth, on the hunt for creatures that live at the outer limits of scientific knowledge. Most of these times, he has brought back old bones, scat samples, tufts of hair, or mouldy pieces of skins for me to analyse. We haven't made a major breakthrough yet, but it is not for lack of effort. Say what you will, but Richard does get out in the field considerably more than most.

I hesitate to call this book Richard's definitive work on cryptozoology, as I am quite certain he's got a good number of books in him yet, but it is quite a read—part autobiography, part cryptozoological encyclopaedia. You may agree or disagree with his conclusions and observations, but I guarantee that once you have read this book, you will have been mightily entertained, and it will have made you think. What more can one ask for in a book these days?

Oh—and should you see, find, or photograph something strange out there, remember to contact Richard (or me)! You never know...

LARS THOMAS
COPENHAGEN, MARCH 2019

CHAPTER ONE

The History of Cryptozoology

'What is the most cunning of all animals?
That which man is yet to see.'
—Indian proverb

This is a book about monsters, real ones. From an early age we are taught that monsters do not exist. This is a lie, pure and simple. Most of us have relegated monsters to the realms of horror movies and nightmares but, by doing this, we do them a disservice. The reality is that monsters walk the earth today. The Oxford English Dictionary has several definitions of a monster. They include 'a large, ugly, and frightening imaginary creature', and 'a thing of extraordinary or daunting size'. Both would sound familiar to most people today. But what makes a monster can depend on how you look at things. Flamingos are beautiful birds. Most of us have seen them at zoos or on wildlife documentaries. They are a favourite amongst birdwatchers, admired for their lovely pink plumage. Most species derive their colouration from the carotenoid proteins in tiny shrimps that they filter from the water with the lamellae in their beaks. Though they look beautiful to us, flamingos must seem horrific to their diminutive prey: as a vast gaping maw fringed with hair-like structures, lunging down to scoop them up.

For the sake of this book, however, I will use the term 'monster' for a cryptid. Cryptids or real monsters fall into several categories. Some, like the Mongolian deathworm, are utterly unknown to modern science; the second category is that of creatures thought to be extinct but that may still thrive in remote parts of the world. The Tasmanian wolf belongs to the latter group. Lastly, we have known creatures who have grown to dimensions far beyond those officially sanctioned by so-called experts. These creatures would include animals like the giant anaconda.

Some cryptozoologists, those who study and search for cryptids, dislike the term 'monster'. I disagree. We would do well to remember the original meaning of the word. It comes from the Latin *monstrum*, meaning revelation. As a cryptozoologist I find this very fitting.

In the days before international travel, before television and before Sir David Attenborough, the average European knew little about the wild animals of the world. Most of their scant knowledge would have been gained, by the few who could read, from bestiaries. These illustrated compendiums of animal lore were popular in Europe throughout the Middle Ages. These books were more concerned with what moral lessons Christians could learn from beasts than the natural histories of the beasts themselves. The dragon, for example, was symbolic of both Satan and pride. Conversely, the unicorn symbolized Christ and his humility. The depictions animals with which we are familiar today such as crocodiles, elephants, and hyenas were so distorted as to look like something else entirely. The artists involved had never seen living specimens of the beasts in question and had only travellers' descriptions to go on. Some of the strange beasts recorded in these works have been shown to be garbled versions of real animals. The manticore, according to the bestiaries, was a beast with red fur, a body like a lion, a man's face, rows of shark's teeth, and a scorpion tail. It was said to dwell in India and be fond of the flesh of men. We know today that the manticore had its origin in the tiger.

As people began to travel more and the globe became more populous, inch by inch superstition was replaced with science. Exotic creatures were captured for menageries springing up across Europe and a new interest in natural history, free from religious allegory, began. With the Renaissance and the dawn of the age of science, many of the ancient monsters were slain by reason. In his 1748 book *An Enquiry Concerning Human Understanding*, the Scottish philosopher David Hume (1711–1776) noted that no amount of evidence could prove the reality of an event that violated the laws of nature. It was more likely that the evidence was wrong. Here came a great stumbling block as even today we do not fully understand the laws of nature. At this time, belief in monsters was waning in Europe, but a few naturalists were still investigating strange creatures.

Baron Georges Cuvier (1769–1832) was a French zoologist, anatomist, and palaeontologist. He established many key concepts such as different rock strata holding different fossils and that animal species could become extinct. In 1812, Cuvier made his greatest mistake. Indeed, it was to become known as his 'Rash Dictum'. He stated that he thought all the world's large animals had already been discovered and there were no more to be found. Just seven years later, one of his former pupils discovered the Malayan tapir (*Tapirus indicus*).

The Prophet of the Kraken

In 1820, a man died of starvation in a cold Parisian gutter. He was given a pauper's burial and the few who attended did so only to snicker at the feeble-minded fool that believed in sea monsters. The man was the French zoologist Pierre Denys de Montfort. Had he lived until 1857, he would have seen his wild stories vindicated. Pierre Denys de Montfort was a malacologist, an expert in molluscs. He was also a scientific heretic, for he dared to research something that the high priests of science deemed to be an old wives' tale: giant cephalopods.

He was born in 1764 and was fascinated by nature from an early age. Sadly, he was of a generation that lost many scientists to history due to the French Revolution and a republic that stupidly—in the words of public prosecutor Antoine Quentin Fouquier-Tinville—did not need scientists. He fared better than most. After serving in the army and after a stint as assistant to the geologist Barthelemy Faujas de Saint-Fond, he became attached to the Jardin des Plantes, the main botanical garden in Paris. For a time, he was a much sought-after scientist, being offered places on a number of expeditions. He travelled to Egypt and Germany to study geology. His gift for languages did not go unnoticed, and he became attached to the Museum of Natural History as a translator.

He investigated the origins of ambergris (the indigestible beaks and claws of cephalopods vomited up by sperm whales like gigantic owl pellets) and became interested in the idea of huge cephalopods. He interviewed American whalers who had settled in France about the evidence for such creatures. One such man, Ben Johnson, told of a monstrous tentacle found in the mouth of a sperm whale they had killed. The tentacle was thirty-five feet long and had been severed at both ends, de Montfort reckoned another ten to twenty feet of it had been lost. It was as thick as a mast with suckers the size of hats. Another man, Reynolds, told of seeing what he thought was a red sea serpent lying next to a whale they had killed. It was discovered to be a massive tentacle forty-five feet long with suckers as large as plates.

Writing in his *Histoire Naturelle Genale et Particulière des Mollusques*, de Montfort classifies two giant cephalopods: the colossal octopus and the kraken octopus, the latter referring to the gigantic squid. He also writes of a votive painting (now long lost) in the Chapel of St Thomas in Brittany. The painting showed a titanic octopus attacking a ship. It was supposedly based on real events involving a ship from a nearby port whilst anchored off Angola. A giant octopus was said to have attacked the ship, wrapping its arms about the rigging and causing the vessel to list dangerously. The crew retaliated with cutlasses and managed to get the monster to relinquish its hold by hacking off some of the arms. The painting was made to commemorate the events, the scared sailors having prayed to St Thomas.

Such huge creatures had been mentioned by Louis Marie Joseph O'Hier, Comte de Grandpre, in his book *Voyage a la Côte Occidentale d'Afrique* written between 1786 and 1787. The natives of that region told him that a giant octopus known as Ambazombi would often attack their boats and canoes, dragging them to the bottom of the sea. They believed the monster to be an evil spirit. The name Ambazombi may be linked to Nizambi, the supreme god of the Bakongo people of Angola. The term 'zombie' is a corruption of Nizambi, whose priests entered a trance-like state before their god entered their bodies.

Captain Jean-Magnus Dens, a Danish man and former employee of the Gothenburg Company, now retired to Dunkirk, told a similar story to the Comte. He had once been becalmed off the coast of West Africa and took advantage of the situation to scrape barnacles off the side of the ship. Men were lowered by ropes whilst sitting on planks. As they worked, a huge cephalopod rose from the water and wrapped tentacles around two of the men, dragging them under. Another arm coiled about a third man who clung to the rigging. His shipmates managed to save him by hacking off the monster's writhing member. The unfortunate man later died of shock. The captain informed de Montfort that the portion severed was twenty-five feet long, and the whole arm had been thirty-five to forty feet long. It tapered to a point and was covered with suckers. The captain felt that if the monster had attached all its tentacles onto the ship, it would have capsized.

Another captain by the name of Anderson told de Montfort of finding two huge tentacles, still connected by part of the mantle, on some rocks near Bergen, Norway. They were so thick he could barely put his arms around them and were roughly twenty-five feet long.

Upon publishing his theories, de Montfort was met with instant hostility in France. Despite being closer to the truth than anyone had imagined, he became a pariah in scientific circles because of his theory. Out of work in scientific agencies, he retreated to the country and wrote books on beekeeping and linguistics. Returning penniless to Paris, he scratched the most meagre of livings identifying shells for naturalists and collectors. He became a wretched, ragged figure and finally a destitute alcoholic. He was found dead of starvation in 1820, a pitiful end for a man once associated with the most august scientific institutes in Paris. Pierre Denys de Montfort has been largely forgotten. Despite having created twenty-five genera still in use today, de Montfort's career has barely merited a footnote whilst his better-off contemporaries, such as Georges Cuvier, are still celebrated today.

Gallingly, after his death, de Montfort was proved correct when parts of a gigantic squid began to fall into the hands of scientists. Danish zoologist Professor Johannes Japetus Smith Steenstrup published the first scientific description of the giant squid. We now know the giant squid (*Architeuthis dux*) can reach sixty feet long.

The Lost Menagerie

As the years went on, more monsters were shown to have a basis in fact. For years, stories had circulated of a savage, hairy giant inhabiting the jungles of Central Africa. Natives called the creature the 'pongo', which was said to kidnap and rape women and tear branches from trees to beat elephants to death. Then, in 1860, French zoologist Paul Belloni Du Chaillu discovered the lowland gorilla (*Gorilla gorilla*). Of course, the stories attached to it were local mythology, but the animal itself did exist. It was not until 1902 that it's relative, the mountain gorilla (*Gorilla gorilla beringe*), was proven to exist when German army officer Friedrich Robert von Beringe shot two specimens in what was then German East Africa. Keep these stories in mind when we examine accounts of other hairy giants in other mountain ranges later in the book. They are now dismissed as the gorilla once was.

The French missionary Father Armand David revived the skin of a black and white bear from a hunter in China in 1869. The creature itself was not seen alive by a westerner, the German zoologist Max Hugo Weigold, until 1916. The beast in question was, of course the giant panda (*Ailuropoda melanoleuca*) a creature now instantly recognisable.

In the same forests that the lowland gorilla called home, there were strange stories of a striped, donkey-like beast called the *atti*. Welsh explorer Henry Morton Stanley heard of this beast whilst in the Congo. Sometime later, the British Governor of Uganda, Sir Harry Johnson, rescued some pygmies in the Congo who had been abducted by a showman. When he returned them home, the grateful pygmies showed Johnson the tracks of the *atti*. Johnson was expecting the animal to be a forest zebra but instead found a cloven-hoofed spoor. He later obtained a skull and skin of one of the beasts. In 1901, English zoologist Philip Lutley Sclater identified it as a species of short-neck giraffe, naming it the Okapi (*Okapia johnstoni*).

Just three years later, British intelligence officer Colonel Richard Meinertzhagen shot a gigantic wild pig in Kenya. It turned out to be the first specimen of the world's largest wild pig, the giant forest hog (*Hylochoerus meinertzhageni*), formally known only in native rumour.

Another traveller's tale was that of a dangerous dragon-like beast said to haunt several remote islands in an Indonesian archipelago. It was said that The Sultan of Sumbawa, a neighbouring island in the chain, would banish wrongdoers, rivals, and other undesirables on these islands to be devoured by the beasts. Pearl fishermen brought back hair-raising stories of encounters with the monsters. The beasts were said to have razor-sharp teeth, claws, and a taste for human flesh.

In a story worthy of Edgar Rice-Burroughs, a western aviator crashed on one of the islands and found himself surrounded by 'dragons'. With the curiosity of western science piqued, an expedition was led by the wonderfully named Governor of Flores, J.K.H. van Setyn van Hensbroek, in 1912. He secured a specimen of what is now known as the Komodo dragon (*Varanus komodoensis*). Growing to over ten feet long and tipping the scales at one hundred and fifty pounds, the Komodo dragon is the largest known lizard and the largest venomous creature, producing a hemotoxin that acts as an anticoagulant in the blood of victims. It kills with razor teeth and a venomous saliva, taking down deer, water buffalo, wild boar, and even humans. The venom hinders the clotting of blood, causing the victim to weaken through blood loss. Large as they are, there are reports of death-dealing lizards much bigger and we will be meeting these in the next chapter.

As the twentieth century continued, more large animals were discovered. On December 22, 1938, Marjorie Courtney-Latimer, the curator of a small museum on the South African coast, was examining the catch from a local trawler. She would often pick up specimens for the museum from the catch of fishermen. Amongst the catch that day was a five-foot-long blue fish with a formidable set of teeth. It had a strange, fan-like tail and fins that resembled stubby legs. Realising that this was unlike any fish she had seen before, Marjorie arranged for it to be transported back to the museum and put on ice. She then wrote to her friend Professor J.L.B. Smith, an ichthyologist of great standing at Rhodes University College in Georgetown, including a drawing and description of the fish. The professor did not receive the letter till the 3rd of January, but when he read it, he recorded his thoughts.

...a bomb seemed to burst in my brain and beyond that sketch and the paper of the letter I was looking at a series of fishy creatures...fishes no longer here, fishes that had lived in the dim

*past, ages gone, and of which only often fragmentary remains
in rock are known.*

On February 16, Professor Smith arrived at the museum to examine
the remains and had his suspicions confirmed. What Marjorie had saved
from the fishermen's nets was a fish from an order thought to be extinct
for sixty-five million years, the Crossopterygii or coelacanths. The fish
was given the scientific name of *Latimeria chalumnea* but was known
simply as the coelacanth in popular parlance.

The discovery sent shock waves through the scientific world and it
wasn't until December of 1952 that a second specimen was captured off
the Comoros archipelago off the northeast coast of Madagascar. Since
then, a number have been caught in the area and even filmed alive in the
deep sea of the islands. But the story had a sequel.

In September of 1997, Mark and Arnaz Erdmann were on their
honeymoon on the Indonesian island of Sulawesi. At a market they came
across what they thought looked like a coelacanth that was brown rather
than blue. They took some photographs and asked fishermen to try to
catch another one. In 1998, a fisherman called Om Lameh Sonatham
caught a second Indonesian specimen. DNA testing proved that it was
genetically distinct from *Latimeria chalumnea* and was dubbed *Latimeria
menadoensis*. This second species of coelacanth has led to speculation
that other species may lurk undiscovered in the deep seas of the world.

Back on dry land in 1971, Dr Ralph Wetzel of Connecticut discovered
a large pig-like beast in the Gran Chaco region—an area of semi-arid
scrub and sparse trees that covers north Argentina, west Paraguay and
southeastern Bolivia. It turned out that this remarkable beast had hitherto
been only known from fossil remains from ten thousand years ago. The
Chacoan peccary (*Catagonus wagneri*), the largest of all peccaries, had
been living happily unknown to modern science since the last ice age.

Into the sea once more and in November of 1976, a team of researchers
from the Hawaii Laboratory of the Naval Undersea Centre were aboard a
research vessel twenty-six miles northeast of Oahu. Two large parachutes
were dropped overboard as sea anchors. They were lowered to a depth of
500 feet. Upon being hauled in, a huge fish over fourteen feet long and

more than one thousand six hundred pounds was found tangled in one of them. The beast appeared to be some form of shark but was unlike anything seen before. The weird beast was frozen and transported to the National Maritime Fisheries Service. There followed a seven-year examination of the extraordinary find.

The fish was found to be a huge plankton-feeding shark with a wide mouth and many rows of tiny teeth. Given the scientific name of *Megachasma pelagios*, this huge fish is more commonly known as the megamouth shark. Dwelling in deep water, it rises closer to the surface to feed on plankton and jellyfish at night. Fewer than one hundred specimens have been recorded since the species was first discovered.

During the 1990s, several new large animals were discovered in the jungles of Indo-China. In 1992, the Vietnamese Ministry of Forestry sent a team to examine the biodiversity of Vu Quang National Park in north central Vietnam. The team came across three sets of horns and a skull from an unknown species. They resembled the horns of the oryx species found only in Arabia and Africa. On three return visits, twenty more specimens were secured including two skins. DNA analysis proved that the new species was a strange member of Bovidae, the family of animals that contains wild oxen such as water buffalo, bison, and yak, and also a number of large antelopes. The creature was named the Vu Quang ox or Saola (*Pseudoryx nghetinhensis*). Since then, live specimens have been captured but have never lived long in captivity.

The same forests also yielded three new species of muntjac deer.

Coming full circle, December of 2013 saw the discovery of a new species of tapir. *Tapirus kabomani*—the little black tapir was found in the Amazon.

This list is not exhaustive, it is simply a sample to show that large and sometimes spectacular creatures can exist unknown to mankind. Why is it then, that cryptozoology is met with such scorn by mainstream science? After so many new and strange creatures have been discovered, you would think that the lesson would have sunk in, but no, cryptozoology is still rejected as a pseudo-science in many quarters. If one examines the cryptozoology page on Wikipedia, for example, you will find an utterly one-sided diatribe against cryptozoology with zero arguments in its favour. This is the kind of childish and inaccurate bias that gives the public the wrong idea about cryptozoology. Why is there this hatred and fear of cryptozoology in mainstream science? The answer may simply be fear of controversy.

In the Victorian era there was a great period of exploration and discovery. Now most scientific bodies are badly underfunded. Fear of losing sponsorship or looking foolish in the eyes of peers seems to have led to a near-hysterical situation whereupon any reports of large new species are rejected as hoaxes, mistakes, or lies. Mainstream scientists may also fear the idea that there are things still unknown to them. This mindset is counterproductive. With so much habitat loss, pollution, and hunting, some species may become extinct before they are ever formally recognised by science. We know this has happened before. The sea mink (*Neovison macrodon*) of the eastern coast of North America was wiped out before it was recognised. It was only described from fragmentary remains in 1903. It was hunted into extinction by fur traders. What other animals have shared or will share the sea mink's fate?

The Monster Hunters

There have always been a minority of scientists who have been open minded enough to embrace the idea of large, unknown animals on land and in the sea.

Charles Gould (1834–1893) was an English geologist and son of the famous ornithologist, John Gould. He took several expeditions into western Tasmania and named many mountains in the Western Range. Gould also conducted the first geological survey of the island.

In 1886, he wrote a remarkable book called *Mythical Monsters*. In the book he speculates on the literal, zoological existence of supposedly legendary creatures such as dragons, unicorns, sea serpents, and the phoenix. Well researched, written, and making a great deal of sense, *Mythical Monsters* is still very readable and relevant over one hundred and thirty years after it was written.

Anthonie Cornelis Oudemans (1858–1943) was a Dutch Zoologist and director of the Royal Zoological Gardens at The Hague. In 1893 he published *The Great Sea Serpent*, an exhaustive look at reports of sea serpents from around the world. Oudemans gave the beast the scientific name of *Megophias megophias* and believed it to be a gigantic long-necked seal. He was almost certainly wrong in thinking this and indeed in thinking that there was only one type of sea serpent but his book was nevertheless highly influential.

The Russians were the first nation in modern times to treat cryptozoology seriously. In the 1940s and '50s a number of Russian scientists were actively seeking manlike monsters in several mountain ranges in Asia.

Boris Porshnev (1905–1972) was a Russian polymath. A biologist and historian, he was very interested in the evolution of man. He was also highly interested in the possibility that relic hominins, relations of the ancestors of man, could still be living in remote parts of the world. He took a number of expeditions in search of such creatures in the former USSR and elsewhere. In the 1950s, he was instrumental in setting up the Soviet Union's Snowman Commission, a group of scientists based at the Darwin Museum in Moscow and dedicated to researching these creatures. The Commission members included scientists like Professor Pitor Smolin, Prof. A.A. Machkovtsev and Dmitri Bayanov. Porshnev himself wrote a number of books on the subject that remain untranslated. Porshnev and his colleagues believed that the creatures represented a surviving strain of Neanderthal man. Now that we know more about *Homo neanderthalensis* we can safely say this is highly unlikely as the creatures reported in Russia and elsewhere seem far too primitive in both appearance and behaviour. They are likely a far more archaic species.

The Commission itself was officially disbanded after several expeditions failed to turn up a specimen but the group continued to meet in the museum. The Snowman Commission was recently resurrected by Demitri Bayanov and Igor Burtsev.

Another member of the Commission was a French-Russian woman, Doctor Marie-Jeanne Koffmann. Dr Koffmann spent decades in the Caucasus mountains searching for the 'Almasty', as the locals called the wildman. She glimpsed the creature once from a distance herself and interviewed hundreds of witnesses and gathered copious notes on the creature's habits. Her impressive and substantial body of work remains mostly unpublished.

Professor Yenshööbü ovogt Byambyn Rinchen (1905–1977) was one of the founders of modern Mongolian literature and a translator. He worked with Porshnev in the Snowman Commision and collected stories and accounts of the almas, or Mongolian wildman.

Another polymath with an interest in cryptozoology was the German science writer and science historian Willy Ley (1906–1969). As a boy in Berlin his teacher asked him to compose an essay on the question 'What Do I Want to Be When I Am Grown and Why?' Ley responded: 'I want to be an explorer.' When his idea was dismissed as 'silly' in true cryptozoological style, Ley became more determined to follow his own ambition.

Ley later studied zoology, astronomy, physics, and palaeontology at Berlin University. He became a science writer of international repute. Horrified by the rise of the Nazis in 1935, he used company stationery to write a letter that authorized a holiday in London. Packing a few books and clothes, he fled for Britain and lived there before moving to the USA.

As well as writing books on general science and science fiction for such publications as *Astounding Stories* and *Galaxy Science Fiction*, Ley had a special interest in cryptozoology. In two books *The Lungfish and the Unicorn: An Excursion into Romantic Zoology* (1941) and *Exotic Zoology* (1959) he speculates on the existence of dragons, unicorns, the Yeti, sea serpents, and latter-day dinosaurs.

Of all the science Bohemians, it is Doctor Bernard Heuvelmans (1916–2001) who remains the most celebrated. Heuvelmans coined the term cryptozoology and is often called 'the father of cryptozoology'. Born in Le Havre, France and raised in Belgium, Heuvelmans studied zoology at the Free University of Brussels. He gained a doctorate by studying the teeth of the aardvark (*Orycteropus afer*). In 1948, he read a newspaper article by Ivan. T. Sanderson that speculated the continued existence of sauropod dinosaurs in Central Africa. This sparked Heuvelmans' interest in unknown animals.

Over the years, Heuvelmans amassed a huge amount of information on all kinds of mystery beasts from around the world. He wrote many books on the subject, the majority of which remain untranslated. His most famous work is *On the Track of Unknown Animals* (1955) which remains the most comprehensive work on unknown animals on a global scale. He followed up with *In the Wake of Sea Serpents* (1958) in which he postulated a number of large, undiscovered marine species, arguing for some more convincingly than others. In the same year he released *The Kraken and the Colossal Octopus*, his book on giant cephalopods. Other works on saurian monsters in Africa and relic hominids remain only available in French.

A great friend and colleague of Heuvelmans was the Scottish zoologist and explorer Ivan. T. Sanderson (1911–1973). Sanderson was born in Edinburgh, the son of a whisky manufacturer. His father was killed by a rhinoceros in Kenya in 1925 whilst assisting a documentary film crew. Sanderson studied zoology, botany, and ethnology at Cambridge University.

A hands-on explorer, Sanderson travelled the world searching for unusual animals. He claimed to have been attacked by a giant bat known as the kongamato in Cameroon in 1932.

Sanderson wrote many books on his expeditions and on zoology in general. His best-known contribution to the cryptozoological canon was *Abominable Snowmen: Legend Come to Life: the story of Sub-humans on Five Continents from Early Ice Age to Today* (1961).

In the days before David Attenborough, Ivan Sanderson was a wildlife expert on radio and television. He also had his own animal collection, Ivan Sanderson's Jungle Zoo. He died of brain cancer in 1973.

Professor Roy Mackal (1925–2013) was a zoologist attached to the University of Chicago. He took part in a number of expeditions to Loch Ness where he hoped to take samples from the supposed monster with a biopsy dart. Mackal thought the monster may be some huge, unknown species of amphibian and published his theory in a book *The Monster of Loch Ness* (1976). He also made two trips to the Congo in search of the mokele-mbembe, a dinosaur-like beast reputed to haunt the jungle rivers and lakes. He wrote a book on his research called *A Living Dinosaur? In Search of Mokele-Mbembe* (1987).

Richard Greenwell (1942–2005) was an explorer and a research associate at the International Wildlife Museum in Tucson. Hailing from Surrey in England, he lived in South America for six years before moving to Arizona in the USA. He was appointed as a research coordinator for the Office of Arid Land Studies at the University of Arizona. He often worked with Roy Mackal and made many trips around the globe searching for Sasquatch, the mokele-mbembe, and the Chinese Yeti or Yeren.

Dr Grover Krantz (1931–2002) was an anthropologist who taught at Washington State University. His research included the origins of speech and of hunting techniques. He studied a number of fossil hominins including Neanderthal man *Homo erectus* and Sivapithecus. At first, he was a sceptic, giving Sasquatch only a 10 percent chance of existing. However, after years of studying tracks and viewing the gait of the subject in the infamous Patterson-Gilmlin film, he came to the conclusion that the Sasquatch was a real creature. He was the first to notice the dermal ridges on plaster casts of Bigfoot tracks and subsequently fingerprint expert John Berry concluded they represented something real.

Krantz wrote widely on Sasquatch, producing five major books on the subject. He also took part in extensive field research but never saw the creature itself despite finding many tracks. After his death, he left his body to science. His skeleton, together with that of his Irish wolfhound, Clyde, are on display at the Smithsonian National Museum of Natural History.

Jeff Meldrum is a professor of anatomy and anthropology and a Professor of the Department of Anthropology at Idaho State University. One of his major fields of study is the foot morphology and locomotion of primates. After examining a set of fifteen-inch-long tracks in Washington State he became interested in Sasquatch and now has a large collection of track casts. His numerous academic papers include *An Analysis of Bigfoot Prints: Evaluation of Alleged Sasquatch Footprints and Their Inferred Functional Morphology*. He has authored and coauthored several books on the subject but his magnum opus is *Sasquatch: Legend Meets Science* (2007). He continues to be an active researcher.

John Bindernagle (1941–2018) was a Canadian wildlife biologist who tracked Sasquatch from 1963. He found tracks and heard vocalizations. He despaired of the treatment the subject received from mainstream science and said, 'The evidence doesn't get scrutinized objectively. We can't bring the evidence to our colleagues because it's perceived as taboo.'

He often worked with Dr Jeff Meldrum and penned two books, *North America's Great Ape: The Sasquatch* (1998) and *The Discovery of the Sasquatch: Reconciling Culture, History, and Science in the Discovery Process* (2010).

One of the most prolific of modern cryptozoologists is Dr Karl Shuker, hailing from the West Midlands in England. He is a zoologist who studied at both Leeds University and the University of Birmingham. Karl, who writes a long running column for *Fortean Times*, has written dozens of books on cryptozoology, zoology, and the mysteries of nature. All are highly readable and have introduced the public to many lesser-known cryptids. He is also the zoological consultant for Guinness World Records.

Jordi Magranaer was a Spanish zoologist. In 1977, he met Dr Bernard Heuvelmans and became fascinated with his investigations into surviving hominins. He spent fifteen years in Chitral in the north of Pakistan in search of the barmanou, the local name for the wildman. He was murdered in 2002, possibly because of his outspoken defence of the hill tribes against the government.

Professor Bryan Sykes is Professor of Human Genetics of Oxford University. His most famous work is *The Seven Daughters of Eve* (2001) in which he traces the population of Europe back to seven ancestral women via characteristic mutations on the mitochondrial genome. Together with Michel Sartori, the Director of the Museum of Zoology in Lausanne, Switzerland, Professor Sykes instigated The Oxford-Lausanne Collateral Hominid Project. The idea was to bring hard science into the search for manlike monsters. Sykes and Satori invited people to send them supposed hair samples from unknown primates such as the Yeti, Sasquatch, Almasty, and orang-pendek. The results were fascinating, and Sykes recorded them in an excellent work *The Nature of the Beast: The First Genetic Evidence on the Survival of Apemen, Yeti, Bigfoot, and Other Mysterious Creatures into Modern Times* (2015). We shall be taking a closer look at his findings in a later chapter.

So, we can see that many respectable scientists have thrown their weight behind cryptozoology over the years, some even risking their reputation in the eyes of a blinkered opposition.

There was an organisation, The International Society of Cryptozoology, founded in 1982 and whose members included Heuvelmans, Mackel, Greenwell. However, the group was poorly run and organised. Its members paying for newsletters and journals that never arrived. It managed to mount a few expeditions but folded in 1996.

A far better and more durable organisation was founded by cryptozoologist Jonathan Downes in 1992. The Centre for Fortean Zoology is based in Devon, England and now has offices in the USA, Canada, Australia, and New Zealand. The Centre has sent many expeditions all around the world researching cryptids and published a journal *Animals & Men* as well as cryptozoological books via its publishing arm CFZ Press.

I became involved in the 1990s whilst studying zoology at Leeds University. Since then I have become the Zoological Director of the CFZ and have conducted research on five continents.

So, cryptozoology seems to be slowly gaining more respect and credibility; however, it needs more hard science, more financial backing, and more time. The first moving film of a snow leopard (*Panthera uncia*) took around seven years to secure. At least as much patience will be needed for most cryptids. Henry Gee, evolutionary biologist and palaeontologist and the editor of *Nature* magazine put it best when he postulated that it was 'time for cryptozoology to come in from the cold'.

CHAPTER TWO

Here Be Dragons

'You, too, the dragons who shine with
golden brilliance...you move on high with
wings, and chasing herds, you tear apart
massive bulls, constricting them in your
coils. Even the elephant is not safe from
you because of his size; you condemn
every animal to death.'

—Marcus Annaeus Lcanus (AD 39–65)

•

The dragon is the great, great grandfather of all monsters. Before the daemon, before the vampire, before the werewolf, before the giant, before them all, was the original uber-monster: the dragon. The dragon's image has crawled across cave paintings twenty-five thousand years old, dwarfing mammoths. It has slithered across Chinese rock art in Shanxi province, eight thousand years before Christ. It haunted the Sumerians and the Babylonians, was worshipped by the Aztecs and feared by the Celts. In the east, a glittering rain god, in the west, a flame-spewing, maiden-devouring monster. It is found in every culture on earth. Recently, Michael Witzel, a Harvard University linguist and philologist, used phylogenetic analysis of legends to trace back their origins to a far more distant time than anybody had previously thought. In his book *The Origin of the World's Mythologies*, he dates the first dragon legends to forty thousand years ago, at a time when the first modern humans were leaving Africa and heading into Europe and across the world. So, it may be that the origin of dragons lies in Africa. The immortal dragon has its fangs and claws deep in the psyche of mankind. And it is still seen today.

Confucius (551–479 BC), the famed Chinese philosopher, wrote of dragons...

I know how the birds fly, how the fishes swim, how animals run. But there is the Dragon. I cannot tell how it mounts on the winds through the clouds and flies through heaven. Today I have seen the Dragon.

The mysterious Greek poet, Homer, wrote of dragons even further back. He is thought to have lived around 800 BC:

An earth-born Serpent, the accursed terror of the Achaean grove, arises on the mead [the grove of Nemea], and loosely dragging his huge bulk now bears it forwards, now leaves it behind him. A livid gleam is in his eyes, the green spume of foaming poison in his fangs, and a threefold quivering tongue, with three rows of hooked teeth, and a cruel blazonry rises high upon his gilded forehead. The Inachian countrymen held him sacred to the Thunderer [Zeus], who has the guardianship of the place and the scant worship of the woodland altars; and now he glides with trailing coils about the shrines, now grinds the hapless forest oaks and crushes huge ash-trees in his embrace; oft he lies in continuous length from bank to bank across the streams, and the river sundered by his scales swells high. But fiercer now, when all the land is panting at the command of the Ogygian god [Dionysos who has caused a drought] and the Nymphae (Nymphs) are hurrying to the hiding of their dusty beds, he twists his tortuous writhing frame upon the ground, and the fire of his parched venom fills him with a baneful rage.

As we can see, the dragon has a long pedigree and perhaps not without good reason.

Terrestrial Dragons

In 1884, the great Victorian naturalist Charles Gould wrote a fascinating tome entitled *Mythical Monsters—Fact or Fiction*. This book was published only twenty-five years after Darwin's *The Origin of Species* and postulated the zoological existence of legendary creatures. Herein Gould tackled such absentees from the ark as the unicorn, the phoenix, the sea serpent, and, most importantly, the dragon. After tracing draconic history for several chapters, he comes to the following conclusions on the nature of dragons.

> *We may infer that it was a long terrestrial lizard, hibernating, and carnivorous, with the power of constricting with its snake-like body and tail; possibly furnished with wing like expansions of its integument, after the fashion of Draco volans, and capable of occasional progress on its hind legs alone, when excited in attack. It appears to have been protected by armour and projecting spikes, like those of Moloch horridus and Megalania prisca, and was possibly more nearly allied to this last form than to any other which has yet come to our knowledge. Probably it preferred sandy, open country to forest land, its habitat was the high land of Central Asia, and the time of its disappearance about that of the Biblical Deluge discussed in a previous chapter.*

> *Although, it probably, in common with most reptiles, enjoyed frequent bathing or basking in the sun, and when not so engaged, secluded itself under some overhanging terrestrial bank or cavern. The idea of its fondness for swallows, and power of attracting them, mentioned in some traditions, may not impossibly have been derived from these birds hawking around and through its open jaws in pursuit of flies attracted to the viscid humours of its mouth. We know that at the present day a bird, the trochilus of the ancients, freely enters the open mouth of the crocodile, and rids it of parasites affecting its teeth and jaws.*

In a way, Gould came closer to the truth than anyone suspected.

We have already met the Komodo dragon, but it had a prehistoric relation that was far more massive. The Pleistocene Epoch (two million to ten thousand years ago), was typified by the proliferation of megafauna. Animals of huge size lived on all the continents (with the exception of Antarctica). Australia played host to some of the most bizarre and amazing. These included:

- Diprotodon—a semi-aquatic hippo sized wombat.
- Procoptodon—a ten-foot-tall kangaroo.

- Nototherium—a marsupial 'rhino'.
- Palorchestes—that sported an elephantine trunk.
- Genyornis—a giant flightless, omnivorous duck!

To prey upon this surreal cavalcade of antipodean herbivores, equally strange carnivores evolved. The most spectacular reptilian macro-predator must have been the giant varanid *Megalania prisca*. This immense lizard reached lengths of thirty feet—rivalling the largest contemporary crocodiles. Much like a scaled-up Komodo dragon, Megalania preyed on the large herbivores of its time. Like the Komodo dragon, it may have had a venomous bite. The projecting armour mentioned by Gould in fact belonged to a giant tortoise, Meiolania, whose fossil remains had been jumbled up with Megalania's. The Aborigines arrived about fifty-thousand years ago. Sharing their environment with such a monster was daunting indeed, and Megalania—almost certainly a maneater—etched itself onto the aboriginal culture. The vast beast became Mungoon-galli—'The Goanna Bunyip'. The natives believed that Mungoon-galli whipped-up sandstorms with its mighty tail, in much the same way as the Chinese believed that dragons controlled the weather. This is one of the many 'coincidences' to be found when one studies dragonlore worldwide.

Around ten thousand years ago, Australia's climate began to change. It became more arid, the rainforests retreated north, and the inland waterways shrank. At the same time, the humans exploited the new dryness and began to light enormous bush fires. Bush fires are a frequent event in many tropical areas during the dry season, but sustained fires with climate change and human hunting was too much for many of Australia's megafauna, and the giant marsupials and birds disappeared.

With its prey gone, Megalania was presumed to have followed them into extinction. But this may not have been the case. The Komodo dragon survived the extinction of the dwarf elephants it used to hunt by killing smaller creatures. There is no reason that Megalania could not have done the same. It should be noted that, as well as the modestly-sized indigenous fauna, there are many introduced species in Australia. These include the feral Asian water buffalo and dromedary camels. Ralph Molnar, an Australian palaeontologist and Megalania expert, said he examined what may have been an un-fossilised Megalania bone! It is part of an ileum or hip bone, and Molnar thought it looked to be between one hundred and two hundred years old! This puts a whole new light on the subject. An animal believed to have died out ten thousand years ago was possibly still around within living memory.

The Aborigines have always told of encountering giant lizards, but as Australia began to be colonised by white men, they, too, crossed paths with the lord of the outback. One of the first sightings took place in 1890 at the village of Euroa, Victoria. A thirty-foot lizard came lumbering out of the bush causing panic and leaving a trail of king-sized footprints. A posse of forty men, armed with guns and nets, set out with cattle-dogs to trap the monster. The beast had other ideas, however, and vanished into the scrub to be seen nevermore.

In May 1961, three loggers encountered a huge lizard. They were in a remote part of the Wauchop forest (New South Wales). Having marked some trees for felling, the trio sat down to brew tea in a previously cleared area. This place was now covered in rotting wood and the loggers heard the crunching of something large approaching them. Looking up they saw a titanic lizard bearing down on them from an embankment. Fleeing in terror the three locked themselves in their truck and watched horrified as the dragon stalked across the dirt track and back into the forest. All agreed that its length was thirty feet, and that it held its body three feet off the ground.

Perhaps the most important sighting happened in 1979 in the Wattagan mountains (New South Wales), as it involved a professional herpetologist—a scientist who specialises in the study of reptiles! Frank Gorden had taken his four-wheel-drive Land Rover into the mountains to look for tiny lizards known as water skinks. After several unfruitful hours searching, Gorden returned to his vehicle and noticed a large 'log' lying on a six-foot-high bank next to the Land Rover. Gorden couldn't recall this log being there before but thought nothing of it until he turned the ignition, causing the 'log' to rear up on four powerful legs and charge off into the woods. Gorden, who was left in a mild state of shock, estimated it to be twenty-eight to thirty feet long in comparison to his Land Rover.

It is highly appropriate that after failing to find any tiny lizards he found one very big one! When a recognised expert in the field sees an animal like this—close up and with a frame of reference for size—doubts about its existence are seriously eroded.

These giants are usually described as being a mottled grey-green, and as in the sighting above are often mistaken for fallen trees. Perhaps this is a form of camouflage employed by the Megalania in ambush hunting. In another case, also from the Wattagans, two farmers stopped their Land Rover to move a log that was blocking the road. Once again, the dead-wood became animate, transforming into a monster far longer than the twenty-foot road was wide. Fortunately, (as the men were outside the vehicle and approaching the monster in blissful ignorance), it could not have been hungry as it merely sauntered off into the bushes.

Further north, in New South Wales, is one of the largest subtropical rainforests in Australia—the Limpwood Nature Reserve. The reserve is a vast plateau of undisturbed forest. It has sheer escarpments and forms part of the Macpherson Range that runs between New South Wales and Queensland. The northern portion is part of a twenty-million-year-old extinct volcano whose lava core now stands as Mount Warning.

In 1984, the late Peter Sleeman was walking through the forest. He was in a stand of what had been dry eucalyptus that had been logged and had subsequently been taken over by the imported Central American shrub lantana. As he walked round the bush, he clearly heard footsteps following him. Cautious as to who might be following him in such a remote area, he side-tracked down a natural pathway to a small clearing and waited for his stalker to emerge.

To his horror, it was not a person but a giant monitor (large lizard in the genus Varanus) that lumbered around the corner. It resembled the lace monitor (*Varanus verius*) except for its huge size. The lace monitor has a head around six inches long. This giant's head was four times this size, and was held around three feet from the ground. He only observed the head, front legs, and front portion of the body. The animal flickered out its forked tongue and tasted the air. Sleeman realised—with dread—that he was being hunted.

Luckily, it seemed Sleeman's hiding place was good, and the monster went on its way. Sleeman left the area and never returned. Scaling his lizard up four times from a lace monitor, we can estimate a creature between twenty and twenty-six feet long. He passed the story on to the brother of Gary Opit, a respected Australian zoologist with many years field experience, who hosts an environmental programme on 2RN North Coast ABC Radio. Gary and his brother explored the area but found no trace of the dragon.

Another report investigated by Gary was a sighting by a woman near Brunswick Heights at the mouth of the Brunswick River. The lizard was sprawled across the road on Fingle Street and covered the entire width—some twenty feet. It then rose up and walked off into the bushes of the reserve.

New South Wales does not have the monopoly on dragons. Many other areas seem to be inhabited by them. The Nullarbor plain in South Australia is one such place. The plain is riddled with sinkholes and catacombs like a vast piece of stilton cheese. Like the caverns of ancient Europe, these are said to be dragon lairs.

In 1940, an aboriginal family made camp close to the catacombs. They slept outside of their truck, off-road. One of the children wandered off at sunset, and despite a frantic search could not be found. The next morning, a trail of outsized lizard prints and the mark of a long, heavy tail were found leading into one of the caves. It was assumed that a Mungoon-galli had emerged during the night and eaten the child. Needless to say, they did not linger long.

In the same region in 1973, two men driving a Jeep late one night, found their headlights illuminating a weird pedestrian—an immense goanna, six feet high and twice the length of the Jeep. Its skin was described as leathery and grey like that of an elephant.

Just how tough their skin is was demonstrated in an encounter from the Margret River area of the Kimbley region of northeastern Australia in 1982. A stockman was mustering his cattle when, to his amazement, he saw a twenty-foot lizard stalking his charges through some long grass. He raised his rifle and fired two rounds at the dragon from a distance of only one hundred and fifty feet. The animal seemed totally unharmed and wandered back into the bush.

Some think that the Chinese reached Australia centuries ago. In 338 BC, Chinese scholar Shin Tzu wrote of animals kept at The Imperial Zoo in Peking. One description is that of a kangaroo, obviously this must have come from Australia, either directly or via trade in the South Pacific. Missionaries who travelled to China in the sixteenth century were shown a sixth-century map of Australia.

If this is the case, could they have encountered huge lizards that added to their already rich lore of dragons?

Australia cannot claim monopoly on giant lizards either, however. Its close neighbour, New Guinea, also has dragons. In the Second World War, Japanese soldiers caught glimpses of what they described as 'tree-climbing crocodiles' deep in the Papuan jungle. Then, in the summer of 1960, panic broke out on the island, when rumours that people had been killed by twenty-foot-long dragons began to circulate. The monsters were said to breath fire and drink blood. Their victims were left with foot-long claw-marks in their flesh. The scare became so bad that the government authorities moved people in the stricken areas into stockades and offered substantial rewards for the capture of one of the beasts. The reward went unclaimed, the dragons disappeared, and the mystery went unsolved for the next twenty years.

In 1960, Lindsay Green and Fred Kleckhan—two administration agricultural officers—found some skin and a jaw bone of one of the dragons in a village near Kariuku. Today, they would have been able to identify these specimens via DNA analysis, something relatively unheard of back then.

In 1969, David M. Davies, an explorer, was shown Papuan cave paintings showing what looked like a giant lizard standing on its hind legs. His native companions reacted with fear at the picture.

Late in 1978 a specimen was finally filmed in Southern Papua by Jean Becker and Christian Meyer. However, even this could not help determine if this was a new species.

In the mid-1980s, famed British explorer Colonel John Blashford-Snell was told of the 'tree-climbing crocodile'. Locals called it Artrellia and seemed to live in great fear of it. He was told that it stood upright and breathed fire. From the descriptions given to him by an old chief, he sketched an animal looking much like a dinosaur. One story told of a young warrior who, many years ago, was hunting deep in the forest. Feeling weary, he sat down on a log. The 'log' in a now familiar style revealed itself as a dragon. It towered ten feet tall on its hind legs, and possessed toothy, crocodile-like jaws. The man fled back to his village in terror.

Intrigued, the Colonel hit the trail. No less a man than the brother of the Premier of the Western Province told him that an elderly man had died in the Daru hospital after being attacked by a female Artrellia protecting her nest. A village elder also said the creatures could grow to over fifteen feet long, and often stood on their hind legs, lending them a dinosaur-like appearance. They were arboreal, and leapt down onto their prey, which they killed with their huge claws and poisonous bite. Even small specimens were feared. A short time before, a small one had been captured and placed in a wooden cage. It swiftly broke free and killed a large dog before escaping back into the jungle.

The Colonel searched for the dragon himself but had no success. He then offered a cash reward for anyone who could bring him a specimen. Eventually, a village priest shot a small Artrellia. It was identified as *Varanus salvadori*—the Salvadori Dragon, (a previously known species but one that none suspected could grow so large). The Colonel later saw several twelve-foot specimens including one huge individual with a head as large as that of a horse was also seen. Such a vast specimen would be in excess of twenty-three feet.

At fifteen feet, the Salvadori Dragon is the world's longest lizard that is currently accepted by science. However, it is not the largest. This accolade still goes to the Komodo Dragon. Over two-thirds of the Salvadori Dragon's length is taken up by its sinuous tail, whereas the Komodo Dragon's tail takes up only half its entire length and far outweighs its elongated cousin. Yet the possibility remains that specimens of this serpentine dragon grow far larger than currently known and lurk undiscovered in this huge island's ill-explored interior. This idea is supported by a number of accounts.

Robert Grant and David George were exploring the Strachan Island district in 1961 when they encountered a grey-skinned lizard some twenty-six feet long. The creature's neck alone measured three feet.

In 1999, two groups of people spotted a dinosaur-like creature at Lake Murry near Boroko. It was six metres (twenty feet) long, with crocodile-like skin. It had thick hind legs with smaller front limbs and a long tail.

As recently as 2004 in New Britain, an island off New Guinea, a giant lizard over twenty feet long—able to rear up ten feet tall and as thick as a 900-litre water tank—was said to be haunting a mosquito-ridden marsh just outside the provincial capital Kokopo, near the devastated town of Rabaul, which was buried by a volcanic eruption in 1994. The monster killed and ate three dogs. The locals thought it was a dinosaur, and police carrying M-16s and shotguns searched the area but found no trace.

In the swamps of the Sudan, there is said to live a beast known as the lau. Natives describe the beast as forty to a hundred feet long, about as thick as a donkey and yellow. Some descriptions furnish it with a crest or mane. Strangely, it is also said to possess facial-tentacles with which it grabs its prey.

The 1920s explorer and naturalist J. G. Millans interviewed a westerner who firmly believed in the monster. Sergeant Stephens (who was never identified with a first name) told him...

> One Abriahim Mohamed, in the employ of the company (a telegraph company), saw a lau killed near Raub, at a village called Bogga. The man I knew and closely questioned. He always repeated the same description of the monstrous reptile. More recently, one was killed by some Shilluks at Koro-a-ta beyond Jebel-Zeraf (Addar Swamps). I obtained some of the neck bones of this example from a Shilluk who was wearing them as a charm. These I sent to Deputy-Governor Jackson (now of Dongola province), who in turn sent them to the British Museum for identification, but no satisfactory explanation was given, nor was it suggested what species of snake they could belong to.

Abrahim's story of the size and shape of the great reptile was corroborated by one Rabha Ringbi, a Nian-Niam from the neighbourhood of Wau in the Bahr-el-Ghazal, who had seen a similar monster killed in swamps near that place:

> Dinkas, living at Kilo (a telegraph station on the Zeraf), told me that the lau frequents the great swamp in the neighbourhood of that station and they occasionally hear it's loud booming cry at night.

> A short time ago I met a Belgian administrator at Rejaf. He had just come back from the Congo, and said he was convinced of the existence of the lau as he had seen one of these great serpents in a swamp and fired at it several times, but his bullets had no effect. He also stated that the monster made a huge trail in the swamp as it moved into deeper water.

Another intriguing piece of evidence was photographed by Captain William Hutchins and published in the magazine *Discovery*. This was a wooden ritual-mask of the beast. When Hitchins questioned Meshengu She Gunda, the native singer and sculptor who made the mask as to the beast's existence the African replied philosophically:

> I might have said, as a young man, when I was ignorant, that there was no such thing as a motor car. I had never seen or even heard of one then. But there is your motor car in the sight of my eyes and I have sat on its chairs and heard its bowels digest inside it. It is thus of the lau.

Latter day dragons lurk in the New World too.

Inhabitants of the back woods in rural Arkansas have folklore of a twenty-foot lizard called a gowrow. The monster sports huge tusks and lurks in caves and under rock ledges. In 1897, William Miller, a travelling salesman, claimed to have killed a specimen near the town of Marshall. He said that he had sent the body to the Smithsonian Institution. Unsurprisingly, however, there is no record of its arrival.

In the 1930s, a gowrow was said to be lairing in Devil's Hole—a deep, wide-mouthed cave on the estate of one J.E. Rhodes. After hearing a loud hissing coming from the cave, Rhodes bravely investigated. He was lowered via a rope two hundred feet below the cave opening, until the narrowing passage prevented further descent. The experiment was repeated using a large stone instead of Rhodes. It was lowered to the same depth as before, and a violent hissing was heard. When the rope was pulled up again, the stone was missing, and the rope had been bitten clean through.

Italy seems to be a Mecca for odd reptiles as well. In December 1933, a snake-like animal attacked farmers in Syracuse, Sicily. It was later hunted down and killed by peasants. The creature was eleven feet long, but its exact nature will never be known as the frightened farmers burnt the corpse.

Far stranger creatures have been reported from Italy. From the 1930s to the 1980s, creatures that seemed to have absconded from medieval bestiaries had been seen roaming the country. In 1935, an eight-foot 'dragon' was seen prowling the woods in Monterose north of Rome. The old man who reported it claimed that he had seen the thing every ten to fifteen years since he was a boy. Its scaled body was green and gold.

More recently, a bigger beast has been at large. A fifteen-foot-long reptile was reported in 1969 near Forli, northeast of Florence. The terrified witness said:

> It was a huge scaly thing at least fifteen feet long. It walked on thick legs and its breath was searing hot. I ran for my life and it followed me for a couple of hundred yards.

Winged Dragons

An interesting theory was put forwards by the author Peter Dickinson in a remarkable book published in 1979. In *The Flight of Dragons*, Dickinson attempts something not done since Charles Gould's *Mythical Monsters*—to explain fire-breathing, winged dragons as real animals within the known zoological frame-work.

Impressed by the universality of dragon legends, Dickinson believed that they had a basis in fact. The main stumbling block was the sheer size of dragons—animals that after all were supposed to have flown. Looking at medieval reconstructions, he reckoned their weight to be around twenty thousand pounds. In order to be able to fly by the muscular power of its wings, a dragon of this weight would need a wingspan of over six hundred feet—far too massive to be real. And how could an animal possibly breath out fire? These problems seemed insurmountable, until a chance viewing of the crash of the Hindenburg.

...one day I happened to see on television an old newsreel film of the wreck of the airship Hindenburg, and almost in a flash all my ideas changed. As I watched the monstrous shape crumpling and tumbling in fiery fragments, with the smoke clouds swirling above, I said to myself, it flamed and it fell, and my mind made the leap to Jordanus. All the pieces I had been considering shook themselves into a different shape. I saw that the Hindenburg was not just a very big machine which flew—it was a machine which could fly only because it was very big. Other answers slotted into place.

Dragons could fly because most of their bodies were hollow and filled with a lighter-than-air gas.

Dragons needed an enormous body to hold enough gas to provide lift for the total weight of the beast.

Dragons did not need enormous wings, because they used them only for propulsion and manoeuvring.

Dragons breathed fire because they had to. It was a necessary part of their specialised mode of flight.

Dickinson's theory held that dragons evolved from large fast-moving carnivorous dinosaurs like Tyrannosaurus rex. They developed huge, chambered stomachs that they filled with hydrogen gas, thus achieving flight. The hydrogen was formed from a mixture of hydrochloric acid in the gut, and calcium from the bones of their victims, and controlled partial-digestion of their own bone structure. The calcium taken from their own bones was being constantly replaced with a regular intake of limestone. This may explain the dragons legendary love of lairing in caves.

The vast body was filled with this gas, and the animal acted—in essence—as a living dirigible. As any chemist will know, hydrogen mixed with oxygen is highly flammable. This is where the dragon's most famous attribute—its fiery breath—came into play. Dragons needed to breathe fire in order to control their flight. To rise, they filled their gas-bag stomachs, and to descend they burnt off gas by breathing it out—possibly with a chemical catalyst, as fire.

The fiery breath doubtless doubled as a formidable weapon—a punishing jet of flame with which to destroy prey and as a display to other members of its species. A similar weapon is employed by the bombardier beetle (*Brachinus*) that spews a jet of boiling chemicals at its enemies. The two chemical components are produced from different glands, and do not reach such high temperatures until it is outside of the beetle's body.

The wings were formed from the extended ribcage—much like that of the modern lizard *Draco volans*, a small gliding species often called the 'flying dragon'. These were covered with a bat-like membrane and were used in navigating the animal in flight.

Dickinson also believes that he could explain some of the more esoteric aspects of dragon legends. The cult of dragon worship would have sprung up from primitive people's fear of such a terrifying creature. The famous 'dragon hoards' would have been built from offerings made to appease the monsters. Dickinson says that dragons would have used gold as a nesting material as it is non-combustible and fairly soft. Their fondness of virgins may have its genesis in human worship and sacrifice of high-born victims, perhaps born and raised specifically as sacrifices to dragons.

The theory also provides a good reason why there are no known dragon fossils. In life, a thick mucus-lining in the stomach-walls kept the powerful hydrochloric acid needed to produce hydrogen in check. After death, the mucous-lining was no longer generated, and the acid destroyed the animal's body. The creature literally digested itself. Hence no dragon bones and no dragon fossils. It is for this very reason that Dickinson's theory is impossible to prove. In effect, he is hoisted by his own petard, and his wonderful theory must, for the time being, remain just a tantalising possibility.

Sightings of winged dragons still occur in the modern age. Picture a ghostly, ethereal globe of light bobbing and flitting in the inky night sky. In the west this would doubtless be called a UFO, and some would deem it to be an alien spacecraft. To the Namaqua people of Namibia, however, this light would mean something infinitely more terrifying—a latter-day dragon. The Namaqua have been reporting such creatures for decades. Their flying snake is described as being the size of a large python, yellow, speckled with brown. From its cranium two horns sprout, and a pair of bat-like wings grow from behind the head. Strangest of all, a glowing ball of light is said to shine on its forehead. This is strikingly reminiscent of the magickal pearls or jewels that were said to be embedded in the heads of Asian dragons. It would be easy to dismiss this as native folklore, but European settlers have seen them as well. In January 1942, sixteen-year-old Michael Esteruise was tending sheep when something emerged from a cave on top of a nearby hill and launched an attack.

> I heard a sound like wind blowing through a pipe, and suddenly the snake came flying through the air at me...it landed with a thud and I threw myself out of its path. The snake skidded, throwing gravel in all directions. Then it shot up in the air again, passing right over a small tree, and returned to a hill top close by.

Michael had been sent out by his father—the owner of a vast farm in Keetmanshoop—to dispel the native mumbo-jumbo that had been costing him both men and money. All of his farm workers had left after he ignored their stories of a giant flying snake that laired in the mountains where his sheep grazed. He finally deputised his boy to show the ignorant savages the folly of such beliefs. The boy did not return. He was later found unconscious and when he came to, related his dramatic tale. Oddly, he related that the snake smelt of 'burnt brass'. Police and farmers investigated in time to see the winged serpent crawl back into its cave. Lighted sticks of dynamite were hurled in after it. After the explosions, they heard a low moaning for a while that gradually died away.

This incident was investigated by no less an authority than Dr Marjorie Courtenay-Latimer, a woman forever immortalised in the annals of cryptozoology as the discoverer of the coelacanth (*Latimeria chalumnae*)—an archaic fish believed extinct for sixty-five million years. She interviewed the boy who took her to the spot of the attack. She did not see the beast but observed the great furrow that it had made in the dust. She also noticed the lack of small animals such as birds and rats.

In 1988, Professor Roy Mackal, better known for his Congo excursions, investigated fantastic claims on a remote property owned by German settlers. Locals described a massive featherless creature, with a nine metre (thirty foot) wingspan, that glided between two hills about a mile apart at dusk. The thing seemed to have lairs in crevices in the hills. Team members discovered the remains of ostrich carcasses in highly inaccessible areas, and believed that the creatures had preyed on them, before taking their kills back to the nest area. Mackal returned to the USA without having spied the animal, but shortly afterwards, one of his team members got lucky. James Kosi—who had stayed on in Namibia for a while—saw the monster from a distance of around a thousand feet. He described it as a giant glider, black, with white markings.

In the mid-1970s, one of the most intense spates of modern dragon sightings reached its peak. The stage was the Rio Grande Valley, and its principle star a reptilian horror that the press dubbed 'Big Bird'. The area was no desolate wasteland but a cosmopolitan area of the state of Texas. Hosts of gun-toting teenagers roamed the streets hoping to cash in on the money prize being offered for the monster's capture. Ornithologists tried desperately to explain the beast away as some mundane, known species. Pelicans, herons, and owls were some of the lamer theories. Those who had seen the thing knew it was no bird of normal pedigree. Some of the sightings were by multiple witnesses.

In mid-January 1976, two sisters—Libby and Deany Ford—saw a black bird as large as themselves skulking near a pond near Brownsville. Upon looking through a book later, they identified it as a Pteranodon—a giant, crested pterosaur believed extinct for sixty-five million years.

The next sighting was not only a multiple-witness sighting, but it involved three elementary school teachers. On February the 24th, Patricia Bryant, Marsha Dahlberg, and David Rendon were driving to work along an isolated road southwest of San Antonio. They were frightened when a huge winged animal flew low over the road, casting an enormous shadow with its wings. The trio estimated its wingspan to be over six metres (twenty feet). All said it had strange, bony wings. Rendon was particularly struck by this.

> ...it had huge wings, but the wings were very peculiar like, it had a bony structure, you know, like when you hold a bat by the wing tips, like it has bones at the top and in between.

The beast was only as high as a telegraph pole and displayed a prominent crest.

Upon reaching the school, they looked through a stack of encyclopaedias until they found a match—the Pteranodon. However, these flying reptiles were believed to be extinct for over a million years.

Elsewhere, other airborne, reptilian monsters have been seen over America.

In 1783, a silversmith and his daughter were chased off Rattlesnake Hill near Silver Run, Maryland. They described their attacker as 'a fiery dragon with gaping jaws'.

In a letter to the *Fredrick News*—a paper in Maryland—a man who signed himself 'R.B.' claimed he had seen a dragon. At 6.30 one morning in 1883, whilst standing on a hilltop, he saw a monstrous dragon with glaring eyeballs, a wide mouth, and a tongue that hung like flame from its jaws. The creature was rearing and plunging above Catotcin Mountain. It sounds like the same type of creature described above.

In 1873, in the skies over Bonham, Texas, a flying serpent manifested itself. Men working on a farm saw a yellow striped snake, the size of a telegraph pole, floating in the sky. The creature would coil itself up, then lunge forwards as if striking at something. Shortly after, the beast was seen over Fort Scott, Kansas. This may have been the same animal that western historian Mari Sandoz recorded earlier:

> Back in the hard times of 1857–'58 there were stories of a flying serpent that hovered over a Missouri river steamboat slowing for a landing... In the late dusk it was like a great undulating serpent, in and out of the lowering clouds, breathing fire, it seemed, with lighted streaks along the sides.

The action then moved to Darlington County, North Carolina in 1888. Witnesses claimed to have seen a winged reptile fifteen feet long that flew as fast as a hawk and made a hissing noise. Reports also came in from three neighbouring counties.

In 1891, the skies over Crawfordsville, Indiana were haunted by a winged reptile twenty feet long and eight feet wide that swooped down from a hundred feet to ground level on violently flapping wings. It scared those who saw it with searing hot breath. The dragon manifested for two nights.

In a letter printed in *Occult Review* of December 1917 by a man who called himself a 'philosophical aviator', the writer had been told of an encounter between a First World War air pilot and a dragon. Whilst at a considerable height, the pilot had seen a dragon rapidly approaching him. He rapidly descended to avoid the colourful reptile. On reaching earth he said nothing to his colleagues for fear they would think him drunk. This account is suspect because the details are so scant. The narrator and the witness are nameless, and the location is not mentioned. The sighting may have been caused by oxygen deprivation at a great height. Even so, another account that is so alike to this may give weight to the pilot's story.

Again, the *Occult Review* took up the story with a letter from Georges Lajuzan-Vigneau printed in April of 1918. He claimed to have seen a letter in a French newspaper in 1909. The story involved three aviators who encountered a huge bluish dragon, that caught up with their plane, and kept pace with it easily. The men panicked, and began to descend, whereupon the dragon seized one of the trio and flew away with him. A grim story but one with little evidence to back it up.

In October 2001, Stevenson Fisher of Camden, Minnesota claimed to have seen a flying monster with a twenty-four-foot wingspan. The grey creature had leather skin on his wings. Fisher said he could see light through them. It flew close to telegraph wires and the roof of his house. He used these to gauge the monster's size. He saw the same creature (or another of the same species) again that winter.

Cryptozoological researcher Linda. S. Godfrey was contacted by witnesses who claimed to have seen dragons in Wisconsin in 2007. One young man she called 'Jim' (a pseudonym) wrote the following...

On October 7, 2007, in Oconto Falls, Wisconsin, some of my friends and I were at a now closed down arcade. It was the same as always, a few of the best local bands, me and some of my friends joining them on stage and hanging out afterwards. While we were outside on the cold, partly cloudy night, one of the guys that was there from Green Bay said the thought he saw something in the sky. Most people were sceptical, but we just decided to lie down on the grass and on top of vans and trucks and maybe we could see what it was.

After about fifteen minutes of talking and laughter, those emotions changed to surprise and astonishment as we watched a massive white/tan dragon fly over the clouds. We knew it had to be a dragon, because how else would you describe something flying over that was almost silent, larger than a plane, had a tail, bat-like wings, long neck, and a narrow, pointed head and scales?

I remember noticing the scales because they dimly reflected the street lights below. We thought we were all seeing things but five minutes later it flew over again, this time in the opposite direction. The eight people from Green Bay wanted to stay but they had to leave. They hoped to see more back home.

Me and my friend K, however, decided to go to my house and lie in the backyard and watch. My mom joined us, not really believing any of it until as soon as she was about to go back in, another big one flew over the house. If I remember right her words were 'I'm tired and I'm going to bed. I doubt that it even— Holy (exclamation)!'

We saw a few smaller ones after that but I haven't seen them since that night. I believe that they were migrating during that month. Hopefully I will see them again next October. When talking about the incident at school, one of the girls around us claimed to have seen the same thing with her cousin two years earlier.

Linda was later in contact with Jim's mother whom she referred to a 'Janet'. Jim had phoned his mother for a lift and told her what he had seen. Janet thought her son was joking and went to pick him up. She noticed a shadow crossing a parking lot then saw a fire ball, blue with orange around it shooting from east to west in the sky. Jim told her that he and his friends had seen the same thing before the dragon had arrived. Janet then drove her son home.

Linda also spoke to Jim's sister Jill. Jim had talked her into staying up. The pair lay on a trampoline in their back yard looking at the sky. They saw a fire ball shooting east to west followed quickly by one moving west to east. Then they heard a loud screech that caused all the local dogs to bark. Jill said...

> Then as we were watching the sky, coming from the west, from the river to over our yard, we saw what appeared to be—and this is the only thing we could think of to describe it as—a dragon. It was making gliding movements up and down, never flapping its wings, kind of like how a mermaid is... It never flapped it's wings at all. And it looked almost cream. The drawing is showing it as we were looking up at the underside. Its stomach reminded me of the underside of a cow, barrel-chested, and from where we were, it looked as large as a cow. I remember the moon was very bright and full that night. It had a snake-shaped head and a long-pointed tail.

The creature was flying only twenty feet above their two-storey house. After looking at pictures of pterosaurs they saw that the wing structure was very different. The family also saw smaller dragons that they took to be the young of the larger ones. The smaller creatures seemed to be playing by circling each other. She continued her statement.

> I don't care what the scientists say, it was NOT a pterodactyl. I could see it had pearly, pale scales. And the fireball came from its mouth. They were beautiful and flew gracefully. I could see they had four legs, too. They were tucked up underneath them like when a bird flies.

In 2010, the Icelandic volcano Eyjafjallajökull erupted, spewing out titanic clouds of ash into the atmosphere and grounding planes for weeks. One local man watching the eruption saw something flying about the volcano. He realised that what he saw must have been huge due to the great distance at which he was observing it from. At first, he thought it was a plane until he saw the wings flapping. In an email to Lars Thomas, the Danish biologist and cryptozoologist, he said the following...

> *As I see it, there are only two possible explanations, either I was imagining the whole thing, or what I saw was in fact a dragon. I find that very hard to believe, but then again, I don't think I could imagine something like that.*

On March 18, 2012, in the southern part of Fayette County in Pennsylvania, a man was walking his dog in a rural location at about 11.45 p.m. He was in his front yard and away from any lights when his attention was drawn upwards after hearing a whooshing sound coming from overhead.

Flying above him at a distance of about fifty-five feet was a large creature that '*looked like a dragon*'. As the flying creature passed, he was able to get a good look at the strange flying animal. The body was about twenty-two feet long with a wingspan of about eighteen feet wide and looked to be shiny with an almost reflective body with no scales. Its colour was dark, auburn brown. At the end tip of the wings there appeared to be talon-like fingers, about three to four in number. The arms of the wing structure appeared muscular. The wings were quite thick, not like skin. There appeared to be a rear fin on both sides of its body, and the creature displayed an arrowhead-shaped tail. The witness also saw what appeared to be two extended rear legs. The creature had a cone shape around the head that stopped flat on the base of the neck.

The oddest physical feature that the witness mentioned to me was that the mouth and eyes were illuminated with, '*a very ominous orange glow*'. As the creature flew over a tree at the bottom of the yard and moved off in the distance, the fellow heard a deep-throaty sound, similar to the fog horn on a boat. The entire observation lasted about twenty seconds.

Dragons of the East

Unsurprisingly, modern day dragons are also reported from Asia.

On August 8, 1934, a thirty-three-foot-long skeleton was found amongst the reeds in Yingkou, Liaoning Province, North East China. The remains had twenty-eight vertebra and a long-horned skull. It was put on display at a local pier and the public considered it to be the skeleton of a dragon. A local paper, *Sheng-ching Shih-pao*, ran several stories on it. The remains were then kept in a school but seemed to have gone missing during WWII.

In 1944, hundreds of people saw a forty-foot, black, lizard-like beast lying on a sandy beach near Chenjiaweizi Village, Jilin Province. It was covered in scales and had whiskers.

A house-sized, long-necked, scaly, green dragon with formidable teeth has eaten fishermen and livestock in Lake Wembu, Tibet. In the 1940s it supposedly attacked a man in a rowing boat, pulling both man and boat underwater.

Jiao Jiao, general secretary for the area, and two friends saw the beast in the early 1970s. They described it as having an ox-shaped body the size of a room with a long neck. They observed it for several minutes.

In 1980, it was said to have killed and eaten a yak owned by a local Communist Party official whom had tethered the animal beside the lake.

In May of 2003, David Nardiello was teaching English in Nigshimozu High School in the town of Watagh Shinke-Cho, Osaka, Japan. He was cycling home late one night through heavy rain. The torrent had formed a pool in some nearby rice fields. Nardiello saw a white animal emerge from the water and turn to look at him. It had a long neck and snake-like head with black shark eyes and fangs. The body and tail were akin to a lizard whilst the four legs resembled a cat. The animal had leathery, featherless wings. It flew into the air to a height of thirty metres (one hundred feet), and Nardiello, increasingly scared, cycled home as fast as he could.

Later that night, he saw it flying through the sky from his third storey flat. He asked his neighbours if they had seen it, but none had. Some, however, said they had heard weird cries from the fields for a few nights. His co-worker Kato Sensi dubbed it Nekohebitori or 'cat, snake, bird'. Nardiello felt strongly that the animal was a predator and was dangerous.

Another strange case comes from Lake Chini southeast of Pahang in Malaysia. British engineer Arthur Potter, his clerk—Inch Baharuddin bin Lajan—and two labourers named Lajan and Malik, all saw a dragon-like creature in May 1959. Mr Potter claimed that he was in his house-boat on the lake at about 10.30 p.m. when it appeared to rain. The rain ceased, then started again. He said:

> Then we saw the attap roof of the boat being raised and I shouted to Baharuddin to flash his torch on it. The roof was about eight feet from the floor of the boat. When the torch was flashed, we saw a gaping hole in the roof nearly three feet across. And we saw, too, a red eye peering down at us. It was the size of a tennis ball.
>
> We couldn't see much more as the torch was not very powerful and the attap blocked the view. Whatever the thing was, it looked exactly the colour of attap. It then disappeared from view. We all thought it was a huge snake and did not think any more about it until the next morning, when we found the tracks the monster had made where it left the water.
>
> Starting from the water's edge we found tracks eighteen feet wide. It looked as if something slimy had gone over the ground. The tracks continued for only four feet and then stopped. But we picked them up again more than one hundred yards away. They were of the same width but stretched nearly two hundred feet.

The sighting earned Mr Potter the nickname 'Dragonwick'.

In the early 1960s, a team of Royal Air Force frogmen dived in that same lake but found nothing.

Explorer Stewart Wavell visited Lake Chini in the 1960s, and was told of the ancient practice of human sacrifice to the dragon around a great pillar of rock that rose up out of the waters. These were believed to pacify the dragon. One of his guides, Che Yang, told him of a great flood that had come to the town of Pekan a couple of years before. Many people had wondered if the dragon was demanding sacrifice, even though the practice had been stopped generations ago. One day a girl out washing her clothes fell from her raft and was drowned. It was noticed that the waters immediately receded. The townsfolk believed the dragon was now satisfied, having taken its victim.

Another dragon-haunted lake in Malaya is Tasek Bera.

In the late 1950s, an officer with the Malayan Police Force went swimming in the lake. After mooring his boat beside Tanjong Keruing—a small headland—he dived into the water. Looking back over his shoulder, he saw a huge neck rearing up over a clump of rassau weed some thirty-six metres (one hundred and twenty feet) away. Two silvery-grey curves showed behind the neck. Panicking, the man swam back to the boat, and paddled away as fast as he could. Looking back one last time, he saw the monster watching his retreat.

The man's commander passed on his account to Stewart Wavell, a producer for Malayan radio, commenting on the man's fine record and reliability. So impressed was Wavell that he travelled to the lake in 1957 in the hope of recording the animal's call which was said to resemble an elephant's trumpeting.

Wavell made camp with his two guides on Tanjong Keruing. Whilst preparing his wire recorder and camera, the monster's cry boomed out across the lake:

A single staccato cry from the middle of the lake... It was a kind of snorting bellow, shrill, strident like a ship's horn, an elephant trumpet, and a sea lion's bark all-in-one.

He switched the recorder on, but the call did not come again.

In 1962, an R.A.F. expedition visited the lake but found no monsters.

Weirdly, the Russian Far East has produced stories, too.

Russian explorer Alexander Remple has been told many accounts about dragon-like creatures in the Russian taiga. Known to the natives as paymurs, they are described as having heads like sheat fish, better known as the wels catfish (*Silurus glanis*) and bodies like crocodiles. It should be noted here that Far Eastern dragons were said to have barbels like those on a catfish's snout.

One man, Anatoly Komandigu, told of three hunters who made their camp by a snow-covered mound in the twilight, and lit a fire. They sat with their backs to the mound and warmed themselves as the fire heated up. Suddenly, they felt the mound at their backs heave. They spun round and saw that the 'mound' was a huge reptile covered in thick grey and black scales. It had short legs and a long tail. Needless to say, the men fled. Three days later they returned for the equipment left behind in the panic. They discovered the remains of an animal, possibly the dragon's prey in the area.

Remple interviewed a seventy-one-year-old man, Vladimir Semyonovich Kuzetsov. Kuzetsov was a seasoned hunter of the Russian taiga. Some years before the Second World War, he stumbled upon what seemed to be some pagan ritual being carried out by taiga nomads. He had spotted a bonfire and heard singing. He quietly approached the clearing and observed a semicircle of people around the fire, chanting songs in a language he did not understand. By the setting sun he saw them performing incomprehensible gestures with their hands. They began to bow down. From the direction of the sunset, he perceived something huge crawling out of the forest. The shape resolved itself into a giant black snake around ten metres (thirty-three feet) long. Kuzetsov believes that he saw the creature's small front legs but cannot be sure. The people raised their voices in a guttural chant and Kuzetsov became afraid. Turning, he fled madly through the trees, not seeing the trail. He forgot how long he ran but his hands and face were covered with cuts when he finally stopped.

How could large reptiles survive in such cold climates? We know that some dinosaurs coped with cool climates, but nothing so extreme as Siberian winters. The huge leatherback turtle or luth (*Dermochelys coriacea*) is a reptile that can survive in cold waters by being gigantothermic. Its size helps it retain its body heat. The leatherback is not an elongated creature like our Siberian dragons. Elongated animals make much less effective gigantotherms than more stubbily-shaped creatures like turtles. Perhaps our dragons hibernate and are active during the brief summer months. But this does not explain the sightings of the monsters in ice and snow. Only until dedicated cryptozoologists journey to eastern Siberia in search of these dragons can we begin to learn the truth.

Dragon legends have many threads forming a rich tapestry of folklore reaching back to prehistoric times. It seems that at least some of these stories are based on encounters with several types of massive reptiles unknown to modern day science. It is interesting to note that the dragon in ancient civilizations was more closely linked with the element of water than that of fire. In the next two chapters, we shall look at water-dwelling cryptids.

CHAPTER THREE

Monsters of Lochs and Lakes

'Ride the snake, ride the snake
To the lake, the ancient lake, baby
The snake, he's long, seven miles
Ride the snake
He is old, and his skin is cold.'

—Jim Morrison, 'The End'
(The Doors LP, 1967)

•

It is only natural that we fear deep water. It is an alien environment in which even the finest divers and swimmers are out of their element. We cannot breathe in it. We are slow in it, it restricts our vision and makes us vulnerable. It is not surprising, then, that many deep lakes around the world are said to harbour monsters.

Loch Ness and Other Monster-Haunted Lakes in Scotland

Gouged out by glaciers in the ice age, Loch Ness runs like a livid scar across the Great Glen of Scotland. Twenty-three miles long, a mile wide, and over eight hundred feet deep, the loch is a fitting abode for a monster. It is said that the entire population of the world—every man, woman, and child—could fit into the loch three times over, and still there would be room for more. The waters are stained inky black by masses of peat washed down from the surrounding hills. This reduces underwater visibility to only a few feet. The cool temperature (42 degrees Fahrenheit), and treacherous undercurrents, add to the loch's dark reputation.

Legends of something odd in the waters date back to at least 565 AD, when the oft-repeated story of Saint Columba was said to have occurred. The Irish saint was in Scotland converting the Picts to Christianity, when he met a water monster said to have bitten a man to death in the River

Ness (not the loch). He overcame the brute with his holy powers and since then, according to legend, the monster has been harmless.

In medieval times, the creatures were known as water horses or kelpies and were believed to drag down and devour humans. Many locals saw a massive creature in the waters over the years, but few reports travelled far. The monster did not achieve true fame in the outside world until 1933, shortly after a lochside road had been built, making the remote area more accessible. A young couple called the MacKays were motoring along the northern shore when they saw the water churning, and observed the two-humped back of a huge animal disporting itself. Shortly afterwards, the *Inverness Courier* dubbed it a monster, and the phrase 'Loch Ness Monster' was born.

Since then, the monster has had more written about it than any other mystery creature. Enough volumes to fill a library have been penned on the subject. An identikit picture from witness descriptions creates a large-bodied animal with a long neck terminating in a small head. A somewhat shorter tail, and four turtle-like flippers can also be seen. The average length seems to be thirty to forty feet and the colour an 'elephant' grey.

The Loch Ness Monster is, arguably, the most famous monster in the world and Loch Ness the most famous lake. Therein lies a problem. Every person who visits Loch Ness will have the monster in their minds, even if it is on a subconscious level. Anything odd seen on the lake may be transformed into a monster by the witness. The deep, sheer walls of the loch can make boat wakes bounce back across the water and are visible long after the vessel that made them has vanished. These wakes can resemble a series of humps. The loch has poorly studied undercurrents that can pull debris such as logs against the wind giving the illusion of a living creature. The loch is connected to the sea by the River Ness in the east and the Caledonian Canal in the west. Seals have been known to enter the loch. Indeed, some years ago I saw an alleged film of the Loch Ness Monster that quite plainly showed a large, male grey seal (*Halichoerus grypus*).

However, not all sightings can be as easily explained. Many witnesses to the Loch Ness Monster report an extreme level of fear at odds with the public perception and tourist board notion of a shy, harmless beast.

One of the most famous eyewitnesses to the Loch Ness Monster is George Spicer, who saw the thing slither across the road in front of his car on July 22, 1933. He said...

> It was horrible—an abomination. First, we saw an undulating sort of neck, a little thicker than an elephant's trunk. It did not move in the usual reptilian fashion, but, with three arches in its neck, it shot across the road until a ponderous body about four feet high came into view... It looked like a snail with a long neck....

Spicer later called for the loch to be dynamited as he had been terribly scared.

Miss Greta Finally, an Inverness housewife, and her son had experienced this recoiling of the mind over twenty years after. They had gone fishing off Aldourie pier on the north shore on August 20, 1954. Ted Holliday, a veteran monster hunter, later interviewed Mrs Finally. She told of waves breaking against the shore and splashing that had made both her and her son look up. They saw a long-necked creature with rough grey skin. It had two projections from its head that ended in 'little blobs'. She confessed to being paralysed with fear. Her son was so horrified that he gave up fishing altogether. She said that she never wanted to see the monster again, even from behind six-inch steel bars.

Mr Richard Jenkyns and his wife saw the beast from their lochside house on September 30, 1974. They watched a sixty-foot thing on the water for half an hour. Mr Jenkyns later commented...

> I felt the beast was obscene. This feeling of obscenity persists and the whole thing put me in mind of a gigantic stomach with a long writhing gut attached.

If Loch Ness was the only lake said to be inhabited by a monster, then we could write it off. However, there are hundreds of lakes across the world where such things are seen. In fact, in Scotland alone there are around thirty-four such bodies of water. After Loch Ness, Loch Morar is the most famous. Morar is a deep glacial loch much like Loch Ness. At eleven miles long, it is smaller, but its depth of one thousand feet makes it the deepest lake in the UK. Its resident monster has been named 'Morag'—a derivative of the Gaelic mordhobhar meaning 'big water'. In centuries past it was believed that Morag would only show itself if a member of a certain Scottish clan was about to die. Morag could appear as a fair maiden or a great serpent.

So much for it being merely folklore, but in the age of enlightenment many people claim to have seen a monstrous creature in Loch Morar. The loch is second only to Loch Ness in reported sightings. Sightings include that of John MacVarish barman at the Morar Hotel on August 27, 1968:

> I saw this thing coming. I thought it was a man standing in a boat but as it got nearer, I saw it was something coming out of the water. I tried to get up close to it with the outboard out of the water and what I saw was a long neck five or six feet out of the water with a small head on it, dark in colour, coming quite slowly down the loch. When I got to about three hundred yards of it, it turned off into the deep and just settled down slowly into the loch out of sight.

> The neck was about one and a half feet in diameter and tapered up to between ten inches and a foot. I never saw any features, no eyes or anything like that. It was a snake-like head, very small compared to the size of the neck—flattish, a flat type of head. It seemed to have very smooth skin, but at three hundred yards it's difficult to tell. It was very dark, nearly black.

Earlier the same year, Robert Duff got a clearer look at the creature. Duff, a joiner from Edinburgh, was fishing in Meoble Bay on the loch's south shore. It was July 8. The water was about sixteen feet deep and very clear. The bottom was pale, almost white with leaves on it. Lying on the bottom was what Duff described as a giant lizard over six metres (twenty feet) long. The skin was a dirty brown, and he saw three digits on the beast's front limbs. It was motionless and looking up at him with slit-like eyes. Terrified he revved up his motor boat and made his escape.

The most dramatic encounter took place on August 16, 1969. Duncan McDonnell and William Simpson were returning from a trip up the loch.

It was around 9.00 p.m. but still light. McDonnell was at the wheel and the boat was doing seven knots. He writes:

> *I heard a splash or disturbance in the water astern of us. I looked up and saw about twenty yards behind us this creature coming directly after us in our wake. It only took a matter of seconds to catch up with us. It grazed the side of the boat, I am quite certain this was unintentional. When it struck the boat seemed to come to a halt or at least slow down. I grabbed the oar and was attempting to fend it off, my one fear being that if it got under the boat it might capsize it.*

Simpson wrote...

> *As we were sailing down the loch in my boat, we were suddenly disturbed and frightened by a thing that surfaced behind us. We watched it catch us up then bump into the side of the boat, the impact sent a kettle of water I was heating onto the floor. I ran into the cabin to turn the gas off as the water had put the flame out. Then I came out of the cabin to see my mate trying to fend the beast off with an oar. To me he was wasting his time. Then when I saw the oar break, I grabbed my rifle and, quickly putting a bullet in it, fired in the direction of the beast. Then I watched it slowly sink away from the boat and that was the last I saw of it.*

Neither of the men seemed to think that the bullet had any effect on the monster. They estimated it to be nine metres (thirty feet). The skin was rough and dirty brown. It had three humps that protruded eighteen inches out of the water. McDonnell thought they may have been undulations rather than humps. McDonnell reported seeing a snake-like head, a foot across, held eighteen inches out of the water.

Loch Shiel is seventeen miles long, a mile wide and four hundred and twenty feet deep. Its monster has been unsurprisingly named 'Sheila'. Father Cyril Dieckhoff of the Benedictine Abbey at Fort Augustus collected many accounts of the creature and was planning to write a book. Sadly, he died in 1970 leaving his work unpublished.

One of the earliest reports is from 1905. An old man called Ian Crookback and two boys watched a three-humped creature through a telescope as it crossed the loch opposite Gasgan in a mail-steamer.

Another report is of a long-necked monster with a wide mouth and seven humps. It was observed through a telescope by Ronald McLeod as it emerged from the water near Sandy Point in 1926.

Loch Oich lies directly south of Loch Ness. In 1936, a headmaster and member of the Camberwell Borough Council, his son, and two friends, saw the loch's monster whilst boating in fairly shallow water. Two humps—like the coils of a snake—emerged only a few metres from the boat. The coils were three feet long, three feet high, and three feet apart from each other. Then a dog-like head emerged. They watched the creature rise and dive several times.

Loch Lochy is Scotland's third deepest after Morar and Ness. On September 30, 1975, at 2.00 p.m., Mr and Mrs Sargent and their two children were driving along the south shore road. As they turned the corner by the Corriegour Hotel they saw a twenty-foot black hump gliding through the water, creating a wave. Mr Sargent slowed his car and his wife fumbled with her camera as the great wash hit the shore. Sadly, the hump disappeared before she could get a shot. Meanwhile, Mr Sargent stopped the van further up the road, and saw a smaller hump following the first. Mrs Sargent did not notice this second hump. The whole sighting lasted about two minutes. Mrs Sargent was visibly shaking, and the children were shouting excitedly.

In 1996, a twelve-foot dark-coloured animal was seen swimming in the loch. It had a curved head and three humps. The creature swam in circles whilst being observed by staff and guests at the Corriegour Hotel. One witness, Catronia Allen, an Aberdeen University Psychology student, observed it through binoculars and said that it was not an otter, dolphin, porpoise, or a seal.

The following year, a six-man expedition, including Loch Ness witness and researcher Garry Campbell and diver Cameron Turner, conducted a sonar sweep of Lochy. Near the centre of the loch they picked up an unidentified reading indicating a twenty-foot object swimming in the water.

The Serpents Saint Patrick Did Not Banish

If Saint Patrick supposedly cast all serpents out of Ireland, he did a particularly shoddy job. The waterways of the Emerald Isle are seething with serpents. In the west of Ireland is a network of interconnected lakes. It is truly amazing that such a wilderness—like something from the Siberian wastes—exists in western Europe. The watery labyrinth is connected to the sea and is the sight of many lake monster encounters. Stories of creatures known as *peistes* or horse eels reach far back into Irish folklore and legend.

In 1954, Miss Georgina Carberry, a librarian from Clifden, and three friends went on a trout fishing trip to Lough Fadda, a small lake that lies in the Derrygimlagh—a bog some thirty square miles, dotted with small loughs connected via streams. What she and her friends saw there would stay with them for the rest of their lives.

They saw a thirty-foot-long creature emerge from behind an island. Carberry stated that it had a shark-like mouth in a head held high out of the water. The body formed two loops or rings as it moved. She also saw a forked tail. She described the beast as 'wormy' with 'movement all over the body'. As the black, coiling thing turned towards them the group fled back to their car and drove away.

Carberry was later interviewed by Ted Holliday. She told him that as she was driving away she found herself watching for the monster. She had the feeling that it had slithered out of the water and was pursuing them. She suffered nightmares about it for weeks after and one of the other witnesses had a mental breakdown. None of the group dared to go back to Lough Fadda for seven years. Even then, they refused to go alone and would never go at night. Think about this for a moment. It takes an abnormal amount of fear to affect four grown women in such a manner.

Lough Mask has had a monster tradition for centuries. The very name reflects this. 'Mask' is derived from the Swedish for 'worm'. A member of the venerable O'Flaherty family was said to have encountered a dragon in the lough. A carving depicting the beast survives today on the O'Flaherty tomb in Oughterad church. On June 16, 1963, at 9.15 a.m., Mr. A.R. Lawrence of Tullamore County Offaly saw the creature. He had his sighting from Inishdura Island at a range of seven hundred and fifty feet.

> *As I stood on the boat-slip looking northwest across the bay, I suddenly saw what appeared to be the head and tail rise of a large fish close to the rocks. A second or so later, the movement was repeated in about the same place. Then there occurred another head and tail rise ahead (northeast) of the first two, followed by the same movement some yards further northeast.*
>
> *I then realised I was watching two humps, one behind the other, moving forwards slowly and regularly across the mouth of the bay. It seemed to me like the back of a very large eel-like fish. I never saw the head or tail, but I would guess the humps were about five or six feet in length, and the distance between them eight to ten feet. The water through which the object travelled was only three or four feet deep. It disappeared from my sight when a point on the island blocked my view.*

Irish lake monsters have been seen slithering across the land on several occasions as well. In the 1890s, a vast eel-like creature was trapped in a culvert between Lough Crolan and Lough Derrylea in County Galway. The creature was said to have rotted away into a foul-smelling slime. F. W. Holiday and Roy Mackal visited the site in 1968. Measuring the culvert, they deduced the creature must have been eighteen inches in diameter.

An almost identical event occurred shortly after, at Ballynahich Lake four miles to the north. A thirty-foot creature—as thick as a horse—became jammed in under the bridge that spans the Ballynahich river close to the castle. A local blacksmith, Patrick Connelly from Cashel, forged a huge barbed spear to slay the beast. This specimen was more fortunate than its comrade. A flood washed it clear of the bridge.

At the turn of the nineteenth century, a Connermara woman was terrified by a horse-eel she met on land. Whilst tending to her turf on a bog next to Lough Auna she heard a commotion in the water. To her horror, a horse-eel slithered out of the water and came right up next to her. The woman ran for her life. She said it looked like horse in front and had an elongated eel-like rear.

Sometime later, a boy and his mother were stacking peat by Lough Auna when they both saw the monster. It looked like a huge eel with a bristly mane and sported three or four humps. It was between nine and twelve metres (thirty to forty) feet long.

On May 14, 1980 a retired Netherlands Royal Air Force officer, Commander Kort, was holding a barbecue at his lochside cottage. He and several of his guests watched a five-foot-long black hump, one foot tall, glide across the lough.

The Lindorms of Scandinavia

Away from the British Isles we find a mass of lake monsters in Scandinavia. The Scandinavia lindorms, legendary serpents believed to have started life as ordinary snakes but over the years grew massive in size. As they grew, they entered deep lakes to become lake monsters. Eventually they became so massive that they headed out to sea and became sea monsters.

In 1555, the exiled Archbishop of Uppsala, Olaus Magnus, then residing in Rome, published his book *Historia de gentibus sertentrionalibus*. This dealt with a history of the serpentine monsters of Scandinavian waters. He writes:

> There is also another serpent of incredible magnitude in a town called Moos (i.e. Lake Mjosa), of the Diocess of Hammer, which as a comet portends change in all the world, so, that portends a change in the kingdom of Norway, as was seen Anno 1522, that lifts himself high above the waters, and rolls himself round like a sphere. This serpent was thought to be fifty cubits long by conjecture, by sight afar off: there followed the banishment of King Christiernus, and the great persecution of the Bishops; and it shew also the destruction of the Country.

In 1636, the cleric Nicolas Gramis recorded that a serpent that dwelt in the Mjos and Branz rivers in Norway had left home and crawled across the surrounding fields. It was said to look like the long mast of a ship, and it knocked over trees and huts that stood in its way.

The most famous of these north European beasts lives in Lake Storjsson in Sweden. This is Scandinavia's answer to Loch Ness. On Forson Island there is a huge megalith upon which a lindorm, swallowing its own tail like the ourobourus, is carved. Along its coils are written Viking runes. A legend tells that a lindorm is bound to the lake until the runes are deciphered. Deciphered or not, a lindorm is resident in Lake Storjsson.

Its first appearance in modern times was in 1820 when a farmer fishing on the lake claimed to have been followed by a huge animal. Another man, Aron Anderson, saw the beast sunning itself on land. It was grey with a white mane on its neck. It returned casually to the water.

Circa 1855 it was spotted by Paul Anderson and four other men. It moved so fast that it overtook the boat they were rowing. In the same year a young man from Ostersund said it pursued his vessel and that it looked like an upturned boat.

In 1863, Jan Brumee and his family observed a dozen blackish humps whilst crossing the lake. They followed the family boat closely. They attempted to catch the monster but it proved too swift. Jan saw it again eight years later.

I was out on the lake with my family and some friends. A rainstorm came up and we headed to shore. That was when the animal moved right past us. It was swimming at an incredible speed. The creature left a very large wake. I didn't give chase that last time, realising this thing had to be strong to swim that fast. With that sort of power, it could have overturned our boat.

Towards the end of July 1878, Eric Olsson and some colleagues were fishing on Storjsson when they saw what looked like a log in the path of their boat. Alarmed when the log revealed itself as a serpentine animal with a dog-like head, they made for a small island and pulled up until the creature vanished.

The Storjsson lindorms also terrorised a couple, Rolf Larson and Irene Magnusson, on a fishing trip in 1976.

We were about five hundred metres from land and going home when we passed a buoy. Suddenly some waves rocked the boat. Fifty or sixty metres from us between the land and the boat, something moved under the surface. Then it came up to the surface, not with a splash but with smooth waves. The part of the body we saw above the water was not more than twenty to thirty centimetres high and about one metre long, but from the amount of water it displaced we could see it was a large object beneath the surface. I would compare it with an upside-down boat. You only saw the keel of it. We had shut off the motor because we were fishing. The thing was swimming in half circles around our boat.

From the beginning I stood completely still without really understanding what was going on. After some minutes I started the motor, but then Irene, who had been quiet all the time, shouted that I should immediately return to land. She was pale as a corpse. We made off shore and the animal followed for a time. We were doing perhaps ten knots, but the animal kept up with us. We returned to land, and we could see the wake of the creature for four or five minutes the water was so calm.

Storjsson is not the only Scandinavian lake that has a monster. Lake Seljord in Norway is another hotspot for sightings. According to legend, the monster—known as the Seljordsorm—once inhabited a smaller mountain lake. It outgrew this abode and moved overland to Lake Seljord. Eivind Fjodstuft saw it whilst fishing in 1920. He described it as black, fifty to sixty feet long, with a head like a crocodile.

In the summer of 1975, dentist Rolf Langeland started a practice in a small hamlet called Sandnes close to the lake. Three days after his arrival he was driving near the lake with his three children when they saw Seljordsorm. He brought his car to a halt, as five humps broke the surface, churning water and moving at an amazing speed. Langeland estimated the monster to be thirty to fifty metres in length. The size sounds excessive, but he may have seen several specimens swimming in line together.

On Easter, Monday 1977, Ivar Hesmyr and his daughter Solveig were fishing from a small boat with a neighbour's son. Suddenly three glistening humps rose from the water three hundred feet away. Hesmyr estimated the beast was nine metres (thirty feet) long. A serpentine head and neck rose up, and the creature began to swim away at a speed that caused the boat to rock. Solveig began screaming and her father attempted to calm her. When he looked up again the Seljordsorm was gone. Solveig said that the humps disappeared first followed by the head and neck. When they reached shore, Hesmyr swore he would never go back out on the lake again.

Another smaller Norwegian lake with a lindworm in residence is Rommen in southwestern Norway, close to the Swedish border.

In 1929, Astrid Myrvold was fetching water for her mother when she saw something on the shore. In later life she likened it to a big, black, plastic pipe (though these did not exist then). It had a horse-like head and a fin on the tail. Disturbed by her presence, it slithered into the lake, drawing a large wake behind it. Astrid noticed it had protruding ears (or horns). She told her mother but was not believed. Hence, she did not speak about it again until 1976 when a local man said he had also seen the monster.

On the 30th of September 1976, at 1.30 p.m., Asbjorn Holmedal was driving a school bus near the lake. He noticed large waves coming ashore between Bjornoya island and the mainland. He thought it might have been caused by a swimming moose. However, when he stopped the bus and got off to take a closer look, he saw a massive animal in the water. He and fifteen children (aged eight to fifteen), saw it rising out of the water and causing great commotion. It was seven to ten metres (twenty-three to thirty feet) long. It had four-metre-long humps, around one to two metres apart. They protruded a foot above the water. The beast submerged, leaving the lake mirror-calm.

In July 1992, Bjord Bohn was holidaying beside Rommen with her husband and daughter. Looking out of her window she saw, fifteen metres (fifty feet) from shore, a hump resembling an upturned boat. She made a drawing of the creature, which she said was patterned like a crocodile. Both her husband and daughter observed the animal before it dove beneath the waters leaving a commotion in the shallows.

Well over forty other Scandinavian lakes are said to have monsters in them. A whole library could be written on the subject, but for now we must move on elsewhere in our worldwide tour of lake monsters.

Lake Monsters behind the Iron Curtain

The Russian Far East seems to be a truly monster-haunted area and some of the strangest stories of monsters come from this vast wilderness.

In July 1953, a prospecting party led by V. A. Tverdokhelbov travelled to the Sorongnakh Plateau. The party arrived at Lake Vorota on a bright sunny morning. Tverdokhelbov and his assistant, Boris Baskator, observed an object some three hundred metres out on the lake.

At first, they thought it was a floating oil-drum, but soon realised that this was not the case as the object swam closer to shore. The pair climbed a cliff to get a better view. In Tverdokhelbov's own words...

> The animal came closer, and it was possible to see those parts of it that emerged from the water. The breadth of the foreparts of the creature's torso, evidently the head, was as much as two metres. The eyes were set wide apart. The body was approximately ten metres. It was enormous and of a dark grey colour. On the sides of the head could be seen two light coloured patches. On its back was sticking up, to a height of half a metre or so, was what seemed to be some kind of dorsal fin which was narrow and bent backwards. The animal was moving itself forwards in leaps, its upper part appearing at times above the water and then disappearing. When at a distance of one hundred metres from the shore it stopped; it then began to beat the water vigorously, raising a cascade of spray; then it plunged out of sight.

Lake Labynkyr lies on the same plateau. It is a big lake nine miles long and eight hundred feet deep. Despite being in one of the coldest regions on earth, the lake never freezes, maintaining a temperature of two degrees Celsius. Labynykr also has an evil reputation. Locals are convinced that the devil inhabits the lake. Gun dogs that have leapt into the water to retrieve shot ducks have been eaten by the monster. One man told of how the brute pursued his raft. He described a dark grey beast with an enormous mouth. Some reindeer hunters observed the monster coil up out of the water to snatch a passing bird.

Author Gennady Borodulin also recounts a tale from Labynkyr in the 1920s in his book *In a Trip to the Cold Pole*.

> *An Evenk family of nomads followed their reindeer and reached the shore of Lake Labynkyr. They decided to stay overnight on the shore. A five-year-old child went to the bank of a stream which led into the lake while adults were busy. Suddenly the adults heard the boy screaming.*

> *The father and grandfather rushed to the bank. They stopped on the edge of water and saw the child being carried away by an unknown animal to the centre of the lake. It was a dark creature, with a mouth looking like bird's beak. It held the child and moved away with quick rushes, then it dived leaving huge waves and dragged the child under the water.*

> *The granddad swore to take revenge upon the 'devil'. He took a sack made of animal skin, stuffed it with reindeer fur, rags, dry grass and pine trees needles, put a smouldering piece of wood inside. He attached the sack to a huge stone on shore with a rope and then threw the sack far into the waters of the lake.*

> *At night there was noise and splashes and terrible screams of the 'devil'. In the morning the waves brought the huge dead animal, about seven meters long with a huge jaw, almost one-third size of the body, and relatively small legs and fins.*

> *The old man cut the animal's stomach, took out the body of his grandson, and buried him on the bank of the stream. Since then this stream is called 'The Stream of a Child'.*

In 1963, a small expedition visited both of these lakes. Four members observed an object eight hundred metres out on Lake Labynkyr. It emerged and submerged several times. They could not take photographs as the sun was setting. The following year, three teams, each replacing the other in shifts, visited the lakes. The third and final group saw the Labynkyr monster in the latter half of August.

Two expedition members saw a row of three humps one hundred metres from shore. They ran after the humps trying (unsuccessfully) to photograph them. The humps dived and rose together. It was not clear if they were separate animals or parts of one creature.

In 1964, two journalists from the Italian magazine *Epoca* visited Lake Labynkyr whilst travelling to Oymyakon. They were told that some time before a party of men saw a reindeer swim into the lake. The deer vanished and did not resurface. Then a dog swam on and vanished as well. Suddenly, and shrouded in a mist, a vast black monster rose snorting from the lake. One of the observers, apparently a scholar, was convinced that the beast was a dinosaur. The locals flatly refused to take the journalists out onto the lake.

Another story concerns a hunter's dog who swam out into the lake and was eaten by the monster. The grieving hunter constructed a raft out of reindeer skin and filled it with hot coals. He floated the smouldering raft out onto the lake. The monster snatched it and dived. It reappeared shortly making terrible sounds.

In the 1970s, a lame horse belonging to some geologists was attacked in the night by some unknown predator. Alerted by the horse's screams the geologists got up out of their sleeping bags to investigate. They were too late. Something large and powerful had already dragged the horse down into the lake. Locals said that they often found holes in the ice with strange tracks around them.

In 2012, Associate Professor of Biogeography Lyudmila Emeliyanova led an expedition to Lake Labynkyr. She and her team recorded big objects in the lake on sonar. She told the *Siberian Times* the following...

> *It was our fourth or fifth day at the lake when our echo sounding device registered a huge object in the water under our boat. The object was very dense, of homogeneous structure, surely not a fish nor a shoal of fish, and it was above the bottom. I was very surprised but not scared and not shocked, after all we did not see*

this animal, we only registered a strange object in the water. But I can clearly say at the moment, as a scientist, I cannot offer you any explanation of what this object might be.

I can't say we literally found and touched something unusual there but we did register with our echo sounding device several seriously big underwater objects, bigger than a fish, bigger than even a group of fish.

Personally, I really believe that something is going on. As local residents for so many years talk about the same strange creature, it just cannot be simply invented. It means that there really is something. Moreover, I know the people who live around the lake pretty well and they are not liars. As icing on the cake, the ancient stories of our Nesski are much older than that of Nessie, so they can't be influenced by the Scottish sightings.

There are many lakes in Yakutia and around the Indigirka River, hundreds of them, big and small, their shores are more or less populated, but all the talk is about Labynkyr and Vorota lakes, and it has gone on for many dozens of years. It makes us think about it. And these stories about the local monster are older than those about the Loch Ness Monster.

As a scientist I know this is not enough to locate and study some unknown creature. I can put it like this, however. I believe there is a mystery in this lake because there is no smoke without fire.

Several fishermen who visit this lake from time to time say they've experienced the following when fishing from a boat in this lake: during quiet, and not windy, weather, when there were no disturbances in the lake, some strange waves coming from under the water suddenly heavily shake their boats.

'It was as if a big body was moving under the water and producing waves which reached the surface and shook the vessel.'

These stories shook me up, for instance, about a boat which was lifted by something or somebody. Two fishermen were fishing in the middle of the lake in late autumn, they were in a ten-metre-long boat when suddenly the bow began to rise as if somebody was pushing it from under the water.

It was a heavy boat, only a huge and strong animal could do such a thing. The fishermen were struck by fear. They did not see anything, no head, no jaws. Soon the boat went down.

This mysterious and very deep lake still has some secret to tell us.

Lake Chany is virtually unknown in the west but it is a vast expanse of water covering 770 square miles. Its is 57 miles long by 55 miles wide but is fairly shallow at only 23 feet deep with an average depth of only 6 feet. Lake Chany is in the southern part of the province of Novosibirsk Oblast, close to the borders of Kazakhstan in southern Siberia.

A story has unfolded in the lake's waters in recent years that sounds like the plot of a horror movie. A powerful, snake-like monster some thirty feet long has said to have killed and eaten nineteen people. The story broke in the west in 2010. The attacks apparently began around 2007.

The creature involved in the attacks is described as serpentine and huge. One witness, sixty-year-old Vladimir Golishev, was in a boat with a friend when the creature overturned it and dragged his friend away. He told the *Daily Mail*...

I was with my friend some three hundred yards from the shore. He hooked something huge on his bait and stood up to reel it in. But it pulled with such force it overturned the boat. I was in shock—I had never seen anything like it in my life. I pulled off my clothes and swam for the shore, not daring to hope I would make it. He didn't make it and they have found no remains.

In 2007, a twenty-three-year-old special services soldier, Mikhail Doronin, was lost when something capsized his boat. His eighty-year-old grandmother, Nina, was watching from the shore and said that the lake was calm. Her eighty-one-year-old husband, Vladimir, said, 'Something on an awesome scale lives in the lake, but I have never seen it.'

Official figures say that nineteen people have vanished in the lake in the past three years. Locals say the figure is actually much higher and that remains have washed ashore with bite marks showing large teeth.

Fishermen have demanded an official probe, but the authorities passed off the deaths as drownings.

A blurry photograph purporting to show part of the monster was released at the same time as the main story. Like most cryptid pictures, it is far from clear but may show a fin protruding above the water.

Lake Chany is too far north and far too cold for crocodiles. The serpentine description makes the Chany creatures sound like huge eels. Lake Chany is landlocked. It has melted water rivers running into it but none running out. Apparently, eels are unknown in the lake but perch, zander, roach, and pike are present. It should be noted, however, that eels often turn up in landlocked lakes and ponds and are quite capable of travelling miles overland. However, Lake Chany lies beyond its known distribution.

The community is held in terror and demanding an investigation. One wonders if an official investigation will ever take place and if it does, what it will find.

Aquatic Monsters in the Orient

Japan has folklore filled with monsters and has modern day creature sightings in its lakes. Lake Ikeda lies near the city of Kagoshuma on the Island of Kyushu. Several photographs have been taken of the Ikeda monster. The first was in 1978 by Mr Toksiaki Matsuhara. He was a folklorist investigating the stories attached to the lake. On December 16, he saw a whirlpool form in the centre of the lake and watched it through his telescope. Later he saw a pair of creatures together close to Metow-Iwa rocks and photographed them. The shots appear to show the two animals together, their humps creating waves. The monster has been christened 'Issie'.

In 1961, a large-scale search was conducted for a US military jet believed to have crashed in the lake. Sonar equipment used in the search reportedly revealed a large object moving through the water below, and records indicate that divers on the lake floor were nearly attacked by a large, unidentified creature.

Many respectable people have seen Issie, including the president of a construction firm. Yutaka Kawaji and twenty members of his family saw a black-skinned animal showing twenty to thirty feet of its body above the surface. Mr Kawaji pursued the monster in his motor boat. It dove and resurfaced several times.

Iamu Horiouchi says that the thing he saw in the lake was a row of humps sixty-five feet long. He took eight photos of the phenomena on the 4th of October 1978, but none came out clearly.

A mere four hours later, coffee shop owner Hiromi Nakahama watched for nearly four minutes as two massive humps rose and fell in the waters.

Issie has counterparts in other areas of Japan. Also, in Kyushu is Lake Toya, home of a similar entity. On the northern island of Hokkaido lies Lake Kutcharo, home to a monster, which has been called 'Kussie'. Local people have formed a protection society for the animal.

The beast first raised its head in 1973. Forty schoolchildren saw it and drew pictures afterwards. The drawings showed a long-necked animal resembling a plesiosaur.

In 1988, eight men in a boat saw a black and white long-necked creature. They too said it recalled a plesiosaur.

In 1990, some odd wave formations were filmed on the lake, but the footage was inconclusive.

Locals say the creature preys on deer and Ainu legends state that a giant snake is supposed to dwell in the lake.

Lake Tianchi lies on the border between Jilin Province, northeastern China, and Ryanggang Province in northeastern Korea. It is a caldera lake within the volcanic Changbai Mountain. Stories of a monster inhabiting the lake date back to the nineteenth century. The creature is known as *Guaiwu* or 'strange thing' in English.

Also, in the late nineteenth century, four hunters encountered a gold-coloured beast with a long neck and horned head which it swung around before diving. In 1908, Lui Jianfeng, a prefectural commissioner, was appointed by the central government to lead an expedition to the area investigating boundaries. He published a monograph *Brief Records of Rivers and Mountains in Chagbai Mountain Region*. This includes accounts of animals including the monster.

In August of 1980, Lei Jai, deputy chairman of the Chinese Writers Association, together with several friends, saw an elongated, ridge backed beast swimming in the lake. His descriptions were noted in the *Guangming Daily*, a national newspaper.

On September 14, 1993 the tourist administration officer for Liuhe county, Liu Shiwen, and several of his friends, observed a creature with a long, mast-like neck swimming in the lake. The creature moved into the shadows of mountains on the lake and vanished.

A number of films have been shot of the alleged creatures but none have been conclusive.

Lake Lieta lies in the Wahui Mountains in Jiulong County Tibet. A monster lurking in the lake is recorded in Buddhist text held in a local temple. The text dates back one thousand years.

Hong Xianlie, a medic based in the area, observed the creature for sixty seconds in 1998. The next day he returned with a video camera. He shot a sequence of film from nine hundred and eighty-four feet away showing a disturbance on the water with something protruding above the surface. He sent his film to a local paper which in turn engendered public interest.

In June 2004, two farmers resting on the shore saw a huge animal with a crocodile-like head and large eyes. It rose six feet out of the water.

In August 2005, Wang Changsheng, a local official, was painting near the shore of the lake. He saw a huge animal with a crested head writhing on the surface. He estimated it to be sixty-five feet long.

The Baffling Bunyip

Australia has a long legacy of water monsters. The most famous of these is the bunyip. The term itself is probably a corruption of the word banip—a term used in the now nearly dead language of the Victorian Aborigines. Bunyip first appears in print in 1812. The bunyips, however, seem to be distinctly mammalian. They fall into two categories: a short-necked dog or seal-like animal: and a long-necked type with a pointed doggish muzzle. Both kinds are hairy. Many believe that the short-necked bunyip was some kind of pinniped, a seal or sea lion that had swum into fresh water. This happens on a regular basis. The grey seal (*Halichoerus grypus*) has also been recorded in Loch Ness. In 1850, a seal was shot at Conargo in New South Wales one thousand five hundred kilometres from the sea. It's stuffed remains graced the Conargo Hotel for many years. The long-necked variety may have been some kind of aquatic marsupial. Tragically, we shall never know. The long-necked bunyip seems to have died out as there have been no reports of these distinctive beasts for well over a century.

Patagonian Lake Monsters

In South America the lakes and rivers of the jungle lowlands are ruled by the giant anaconda and possibly by even larger serpents. But the deep cold lakes of the Andes hold a different kind of monster.

Lago Nahuel Huapi in Argentina is the best-known monster-haunted lake in South America. The first modern sighting occurred in 1910, when Mr George Garret and his son were navigating a government engineer around the lake. They briefly saw a twenty-foot hump rise six feet from the water. But things didn't really heat up until the 1920s.

In 1922, the director of the Buenos Aires Zoo, Dr Clementi Onelli, received a most intriguing letter from an American prospector called Martin Sheffield. Sheffield was a larger than life character who seemed to have stepped out of a western. Heavy drinking and six-gun toting, he always wore a ten-gallon hat. At the time, Sheffield was searching for gold in the Andean foothills of the Chebut territory in Patagonia. Beside the lake he had come across the spoor of a large animal that had crushed bushes and left deep marks in the ground, disappearing into the lake. In the lake, the prospector claimed to have seen a monster with a swan-like neck and a crocodilian body.

One could have forgiven the good doctor for ignoring the testimony of a such a man as Sheffield, but he had heard other reports from South America himself. Whilst travelling in Patagonia in 1897, he had been told of a monster inhabiting White Lake. A Chilean farmer recounted that he had heard strange noises at night as if some heavy cart were being dragged over the pebble beach. On moonlit nights he had seen a giant beast swimming in the lake raising its reptilian neck high out of the water.

Onelli tried to get funding from a film company but was refused. Undaunted, he raised the three thousand pesos he needed by public subscription. In May 1922 the expedition set off. It was led by zoo superintendent Jose Cihagi and Emillio Frey, an engineer who had been a member of the Chile-Argentine Boundary Commission for twenty years and who had explored the Chebut territory. Carrying elephant guns and dynamite, the expedition travelled by boat, lorry, and horse into the then wild area.

The enterprise caused outrage in some quarters, with the president of the Society for the Protection of Animals demanding that the Minister for the Interior should not grant the expedition permits to hunt the monster, under state law 2786, which governed the hunting of rare animals. After confusion about permits that halted the expedition for a while, they finally reached the lake on October 18.

Borrowing a boat from some Welsh colonists, Frey detonated eleven cartridges of dynamite in the lake to absolutely no avail. Finally, winter forced the expedition to head home.

Back at White Lake, a British man named Barny Dickenson, who had lived by the lake since the late 1940s, was prompted to write to the *Illustrated London News* in 1959 after a series of articles on the Loch Ness Monster had been printed. He often sat by the lake of an evening watching boats go by. On several occasions he had seen an animal he thought was too large to be any known South American beast in the waters. The local Auracanians called the thing Curo, meaning cow hide, on account of its texture. Dickenson said that the local gauchos also reported having seen the monster. It appeared in the evenings as a single huge hump, like an upturned boat with the texture of cowhide.

Today a city of over one hundred thousand souls, San Carlos de Bariloche, has sprung up in the once remote area. The sightings of the monster, now named Nahuelito, continue. Father Maurico Rumboll's mother saw a long neck break the mirror-like surface of the lake on February 18, 1978. It left a large wake on the surface as it advanced.

In 1989, Cristian Muller had set out at 7 a.m. for a day's fishing. Travelling by bus, he glanced out of the window and saw what he at first took to be a boat. Its dark colour was very odd as all the boats he knew on the lake were light in colour. Suddenly the 'boat' submerged, and the bus driver slammed on the brakes. The excited passengers began screaming: 'Nahuelito, Nahuelito!'

On New Year's Day in 1994, Jessica Campbell and several other people saw the monster as a row of humps showing above the surface as it swam.

On another occasion, Paula Jacarbe claimed to have heard the monster breathing. Charlie her brother, the keyboard player in a band, attempted to replicate the sound with a synthesizer.

In 1996, Jessica Campbell saw the monster twice more. At the beach at Peninsula de San Pedro the beast surfaced in front of a whole group of people including Campbell. The watchers noticed leathery skin and cloak-like fins. Nahuelito submerged but rose again forty-five minutes later, directly in line with Campbell who was sitting on some rocks. As the monster swam towards her, she was gripped with terror and fled. Much like Paula Jacarbe, she recalls hearing the monster breathing. She said she would never forget those sounds.

North America: Lake Monster Central

It is in the northern part of North America where we find the most reports of lake monsters. Montana's Flathead Lake has a monster recorded in the legends of the Kalispell Indians. They spoke of a monster with an arrow-proof hide who would snatch up and devour braves. Captain James

Kern was the first non-native person to see the beast. Whilst piloting his steamboat across the lake in 1885, he saw what he took to be another vessel. Soon it revealed itself to be the humped back of a huge animal. Passengers, with typical human care and concern, opened fire on the creature, which promptly dived back into the water.

Another steamboat captain had a run in with the lake's strange inhabitant in 1919. Seeing what he thought was a log, he took evasive action to prevent a collision. The 'log' sprang into life like its kin in other lakes and swam round to the starboard side of the vessel where fifty passengers saw it, though this time no one took pot shots!

The Flathead Lake monster has been credited with ripping up fishermen's nets for years.

Members of the Zeigler family were alarmed by unusually large waves lapping over the dock by their lakeside home. Upon investigation they discovered a horse-headed monster scratching itself against the pilings. Mr Zeigler rushed inside for a rifle. Luckily (more than likely for him), the monster had swum out of range by the time he returned.

Another lakeside resident, Ronald Nixon, saw the monster in 1963. It was twenty-five feet long and threw up a two-foot head wave. It showed no signs of a dorsal fin and did not seem to be a fish.

In the summer of 1969 or 1970, Kay Grice, her sister, her niece, and a friend, were in a boat off Arrowhead Island when they noticed a series of 'v' shaped wakes in the water. They followed the trail in their boat, watching as it smoothed-out, then formed again. Suddenly a large creature rose up fifty feet behind the boat. It had a long neck and reminded them of illustrations of the Loch Ness Monster.

In 1975, a twenty-three-year-old girl from Indiana was on the lake in a sailboat belonging to the Olsen family. She returned to the dock pale and shaking and told the Olsens that a long, brown, snake-like thing had swum under the boat.

The Olsen sisters saw the animal for themselves in 1982. Twenty feet of its body was showing in two humps. The head was oval shaped and larger than an American football.

Major George Cote and his son Neal have seen the Flathead monster twice. The first sighting was in May 1985 as they trolled for trout in Mackinaw Bay:

> *We saw a large object surfacing and diving off the north point of the bay. At first, we thought it must be one or more SCUBA divers. We approached the thing slowly. As we got closer, we could see it was chasing large squawfish in the shallows. At one point it raised its head out of the water and appeared to be looking at us.*

> *When we got within sixty metres of it, we realised it was like nothing we'd ever seen. The thing was big: as long as a telephone pole and twice as large in diameter. The skin of the creature was smooth and coal black; it had the perfect head of a serpent. There were four to six humps sticking out of the water. It stopped four hundred metres out from the bay, looked back, and dove under the waves.*

They saw a similar animal in 1987 off lakeside on the west shore. They estimated its speed at an impressive one hundred knots. They had a perfect view of its head, tail, and body as it swam towards Caroline Point.

In July of 1985, George and Elna Darrow observed it from their sail boat three hundred feet offshore. It manifested as three dark loops thirty feet long.

One US lake monster stands pre-eminent amongst its kin. A long neck above the others in terms of the number of sightings and its worldwide fame, the monster is, of course, 'Champ' of Lake Champlain. This one hundred and nine-mile-long lake lies between upstate New York and Vermont. It just crosses the border into Canada at its northeastern tip. It was formed around ten thousand two hundred years ago as the Champlain Sea, an arm of the Atlantic, receded and glacial meltwaters formed a freshwater lake.

The Iroquois Indians, who lived on the lake's western shore believed in a great horned serpent that lived in the body of water they called 'Petoubouque' (the waters that lie between).

Since 1819, there have been hundreds of reported sightings of 'Champ'. Let us look at a few of the more dramatic.

Tommy Heinrich, a fifteen-year-old high school student from Burlington, had a breathtaking encounter whilst fly-fishing from a boat on the lake in the 1970s. A snake-like wriggling form, twenty-five to forty-five feet long, humped up from the depths, water cascading off its rust coloured skin. A horned, equine head swung round to look at him before the leviathan dove back to whence it came. Fishing forgotten, he made it to shore in record time.

Fred Shanafelt and Morris Lucia were two experienced scuba divers from New York. No amount of experience could prepare them for their run in with 'Champ', however. It occurred whilst they were on holiday near Moquam Bay in May 1972.

After breakfast, the men were scuba diving from a small boat. Upon surfacing, Shanafelt saw Lucia frantically signalling to him.

> From his actions I knew we were in some kind of danger. After he saw I understood the situation was dangerous, we both started back to shore. I surfaced about ten feet out, looking back to see what had given Morry such a fright. That's when I saw this thing that couldn't have been anything other than a sea serpent. Christ! That sounds crazy but there's no other way to describe it.

Lucia described Champ thus...

> It was hard to judge how long the body was. I estimated around forty feet, and Fred believes it was more like forty-five to fifty. We agreed the head looked about like a horse. It was slightly rounded in appearance, a sort of mushroom grey colour. The head sat on a long round neck that was dark brown or black. The neck rose up about eight feet out of the water at the highest point.

The men swam to the beach and clambered up. The beast swam closer, apparently appearing more curious than threatening. It cocked its head whilst regarding them. The monster and the men stared at each other for two minutes before Champ swam away and dived.

The summer of 1977 saw the taking of the best still photograph of a lake monster. Sandra Mansi and her husband Tony had taken their children for a day out beside the lake. The happy atmosphere was shattered. About one hundred and sixty-two feet from shore, something began to emerge. Sandra thought it was a large fish until she saw an elongated neck atop a large body rise. The Masnsis thought it looked like a 'dinosaur'. The monster swivelled its head and neck to look around it. Tony ran to get the camera from the car as the monster moved in closer. They estimated the visible length to be six metres (twenty feet). Thinking it might be aggressive, the Mansis moved their children away from the water. Tony handed his wife the camera, as he bustled the kids back towards the car. Sandra shot one picture of the monster before she too fled and the family sped away in the car.

The picture appears to show a snake-necked animal looking backwards over a substantial body. It is of startling clarity in comparison to most other lake monster pictures. The family were badly frightened by the encounter and forgot at exactly what point in Champlain's vast waterline they had been at.

The Mansis never sought any publicity, and it was not until several years later that they went public, and that was by chance. When her bosses at General Dynamics selected her to take part in a project in Scotland she did not want to go. A co-worker who went in her place said that he wanted to visit Loch Ness as he had an interest in the monster. Sandra told him that he did not have to travel that far to see a monster.

The man did not believe her until she showed him the photograph. Soon news got around, and Sandra went to Washington, DC, to formally have the photo copyrighted. The snap was brought to the attention of Dr Phillip Rines of the State University of New York. Dr Rines passed on the information to long-time Champ Researcher Joseph Zarzynski who sent a print on to Dr George Zug at the Department of Vertebrate Zoology at the Smithsonian Institute. Dr Zug felt that the photograph was probably genuine and sighted the recently discovered fact that sea turtles can travel much farther north and withstand much colder climates than was ever thought possible.

The picture was examined by Dr B Roy Frieden of the Optical Sciences Centre at the University of Arizona. He was confident that it was not a superimposition or montage of any kind, on account of a set of waves being made by the objects itself and separate from the naturally occurring waves on the lake.

Paul H LeBlond of the Department of Oceanography at the University of British Columbia has also studied the picture.

> *Empirical results relating to the appearance of the sea surface to wind speed, and thence to the length of wind waves are used to provide an estimate of the dimensions of 'Champ' as seen in the Mansi photograph. Over the possible ranges of wind speed and fetch, for the lower and upper waterline dimension of 'Champ' range from 4.8m to 17.2m.*

The Mansis' estimate of the monster's size falls close to the lower end of LeBlond's size range findings.

In the 1990s, an expert from Kodak who was highly experienced with the make of the camera Sandra used (Kodak Instamatic), examined the picture for the cryptozoological documentary series *Supernatural*. His conclusions closely matched LeBlond's and Freiden's. They photo is not a montage or superimposition. The size of the object in the picture is between twenty to seventy feet.

The Great Lakes of the eastern USA, and southern Canada, five linked massive bodies of water, are not without serpent reports. Samuel Rafinesque, a biologist who catalogued many North American species wrote of a thirty-five to sixty-foot serpent that had been seen on Lake Erie in July 1817. It was a foot thick, dark mahogany in colour, and had shining eyes.

Over the years there have been a number of sightings. In 1960, Ken Golic was fishing close to a pier in Sandusky at about 11 p.m. He saw a cigar-shaped hump protruding one to one and a half feet out of the water.

Jim Schindler came within six feet of the monster as it swam a foot below the surface close to South Bass Island in 1969. He was unable to guess the monster's length but estimated its width to be two feet.

In 1981, Theresa Kovach reported seeing a 'snake-like reptile so large it could easily capsize a boat'. She observed the thing from her house on the Cedar Point Causeway.

The Lake Erie monster has been named 'South Bay Bessie', continuing the slightly irritating convention of giving monsters 'twee' names. Bessie was seen again in 1983 by Mary Landoll off Rye Beach, Huron. Just before dawn, whilst sitting at her porch, she heard a sound akin to rowing. Looking out over the still lake she saw what looked like an upturned boat. It was greenish brown and twelve to fifteen metres (forty to fifty feet) long. A long neck with a head became visible and she noted both a mouth and eyes.

A serpentine beast with five humps and a flat tail was seen in 1985 by Tony Schill whilst out on a boat with his friends. A similar serpentine form was observed off Lorain Coast Guard Station by Dale Munroe. The three-humped animal was in view for four minutes and was twice the size of his sixteen-foot boat.

Bessie has turned up on sonar as well. Gail Kasner discovered a cigar-shaped object on a fishfinder image graph in 1985. The object was thirty-five feet long and thirty feet below the surface.

Jetski enthusiast Bob Soracco thought he was looking at a line of porpoises on September 3, 1990. He was later informed that there are no such mammals in Lake Erie. The row of grey humps had dived as he drew near them.

The gallant Harold Bricker wanted to get a closer look at Bessie, when the monster rose some one thousand feet from his family's fishing boat. It revealed a snake-like head and thirty-five feet of black body. His son, realising that the monster was bigger than the boat, vetoed his father's idea. The monster submerged, perhaps luckily for the Brickers.

On September 11, 1990 two fire inspectors, Steve Dicks and Jim Johnson, watched the monster from a third storey window. It showed three black humps and remained in sight for three minutes. Several other sightings were reported in the following few days.

Huron Lagoon Marina offered five thousand dollars to anyone who could prove Bessie's existence. The reward is still unclaimed but 1993 was a bumper year for sightings. Charles Douglas and a companion were eight and a half miles out on the lake when they espied what they thought was a log. It was a particularly lively log, as it swam at about thirty-five mph and frequently dived. It appeared to be following their boat. Douglas wanted to get a better look, but his less courageous chum dissuaded him. Cowardly friends are a bane to cryptozoology!

Another acrobatic log was sighted shortly afterwards by charter fishing operators John and Holly Liles and their party. As they were taking the party out to Kelly's Island at four in the afternoon, they sighted a log floating one hundred and eighty feet away on the lake's calm surface. As they drew closer, they realised that the twenty-five-foot log was an animal, the likes of which they had never seen before. It undulated in a series of vertical humps and reminded the witnesses of a Chinese dragon. It was in sight for around fifteen seconds before the boat engine disturbed it and it dived.

Though Lake Erie boasts the most sightings, it does not have a monopoly on the Great Lakes monsters. The year 1977 saw a spectacular sighting from Lake Superior. It was Memorial Day, and Randy L Baun was camping at Presque Isle, north of Ironwood, Michigan. Taking a trail that led east from the campsite, he slid down a steep wooded bank to a small beach. Here he began to eat lunch. Looking out across the lake he saw three undulating black humps about a thousand feet out. The humps swiftly swam towards him, and he became alarmed as he saw a monster he described as looking like 'an anaconda with the girth of a Volkswagen'. He grabbed his 35mm Yashica camera and took a shot, as a horse-like head appeared. It had a large eye and twitching, catfish-like whiskers on the snout.

Baun's photo shows a dark head with a large eye reflecting light. The animal remained in view for thirty seconds, but Baun was frozen with fear. He still has nightmares about being devoured by the Lake Superior dragon and will not swim in deep water. He says there have been many disappearances in the area and believes that the monster may have eaten people.

In 1997, a recreational fisherman claimed to have seen a buck bitten clean in two by something in the lake as it waded in the water. It seems that swimming in Lake Superior may well be considered an extreme sport.

Lake Michigan has its resident monster as well. The *Chicago Tribune* reported that on August 6 and 7, 1867, the crews of several vessels had seen a serpent forty to fifty feet long, and as broad as a barrel. The animal beat the lake's surface into a tempest with its bilobate (fluked) tail.

Joseph Muhlke, a fisherman, saw it whilst a mile out on August 6. Disturbed by a strange grating vocalization, he turned to see a dark oval object like an upturned boat some one thousand three hundred feet away from him. He watched as the object approached, then a head surfaced with clearly visible eyes, and a tail came into view. It was bluish black, with a paler underside. Along its back were a series of bony plates. Muhlke noticed flippers and a fan-like fin on the tail.

Pulling up anchor, Muhlke made for shore. The monster swam off, creating a wake that shunted his boat inshore at a considerable speed. Sightings were recorded for several days afterwards.

Kimberly Poepey Del-Rio was travelling by bus beside Lake Michigan in February 1988. Two stops before her home, she noticed something odd swimming in the water of a cove next to the War Memorial Art Centre. The centre is built on a peninsula reaching out into the lake. Pulling the cord to stop the bus, she ran over and saw a hump—like the top of a car—manoeuvring around large lumps of ice in the cove. It finally swam back out into the lake.

The Centre for Fortean Zoology's Indiana representative, Elizabeth Clem, has unearthed several encounters on Lake Michigan. Witnesses emailed her with their own stories. One of them claimed to have been fishing with a friend on the lake, when something grabbed his line and yanked so violently that their boat capsized. The two fishermen climbed on top of the upturned boat as the water became more disturbed. Then they saw, about one thousand feet from them, a horse-like head on a long neck emerge from the lake. The neck sported a mane. They screamed until some of their other friends rescued them in another boat.

Another witness had a strikingly similar experience. He was fishing in a boat with his brother when the vessel was violently rocked. Not being able to swim, and thinking it was his brother fooling around, he told him to stop rocking the boat. His brother insisted it was not him. Just then the boat was shaken again, and they saw a large black object swim beneath it. When a large head surfaced only twenty feet from the boat, they started the motor and exited the area quickly.

Monsters do not recognise the false boundaries that humans erect between countries. As we cross over into Canada, we discover a country seething with serpent dragons, perhaps more than in any other country.

Literally dozens of bodies of water in Canada are the reputed homes of such creatures. It would be a gargantuan task to look at each of them, so I have selected the ones with the best or most numerous sightings, leaving the greatest of them all for last.

Lake Manitoba in Manitoba is home to a monster known as 'Manipogo'. It was named after the better-known 'Ogopogo' of Lake Okanagan (more on that later). Manitoba is shallow in comparison to most monster-haunted lakes (thirty feet) but it may be that Manipogo is not a full-time resident.

In the summer of 1957, a group of journalists set out to find Manipogo. They came across a cave that had the remains of many small animals in it. They said traces of a heavy snake-like body were visible. None of the journalists laid eyes on the creature, but their Indian guide, Solemn Fleury, claimed to have seen a thirty-foot animal in the lake, that roared at him before diving. The shock caused him to faint. A not-so-brave 'brave' I guess.

Twenty picnicking people were stunned by the appearance of Manipogo in July 1960. One man, A. R. Adams, ran along the shoreline, keeping pace with the monster. It showed three black humps. He saw the top of its flattish head. This mass sighting caused Dr James McLeod, head of the Zoology Department of the University of Manitoba and Siggi Oliver of the Provincial Fish and Game Department to investigate the sighting seriously. They ended up more baffled than when they started, as it was quickly apparent that the creatures in the lake were unlike anything known to science.

Meanwhile, the monsters were not idle. One surfaced just yards from a man and two women rowing off Graves Point, causing them to panic and head for shore. Then, in August 1960, three of the creatures appeared to seventeen people at Manitoba Beach. They resembled huge brown snakes. Two swam side by side, while the third trailed behind. One witness, Thomas Locke, tried to film them, but discovered he had run out of film.

Another in the trinity of 'pogo', namely 'Igopogo', is said to be seen from time to time in Ontario's Lake Simcoe. In 1963, a Presbyterian minister and funeral director Rev L B Willams and Neil Lathangue, together with their families, were boating on the lake. They saw a huge animal coming towards them. It was charcoal coloured and thirty to seventy feet long.

It had a dog-like face, and a neck as thick as a stovepipe. It also sported dorsal fins along the back.

A sonar reading of a large animal was taken in Simcoe in 1983 by William W Skrypets, from the Government Dock and Marina.

Without a doubt, the most famous monster in Canada is 'Ogopogo', the monster of Lake Okanagan. In sheer number of sightings, Ogopogo is second only to the Loch Ness Monster. The lake itself is shaped like a writhing serpent, mouth agape. It lies in British Columbia. It extends some eighty miles from Vernon in the north to Princeton in the south. Its maximum depth is one thousand feet. It is impressive, but far from the largest of the North American monster lakes. But size is not everything, for no other lake on the continent can hold a candle to Okanagan.

Native Americans had a great fear of the monster and believed that it could only be placated by sacrifice. If crossing the lake, they would carry a chicken or some other small animal. These they tossed into the water for the monster to eat. The beast was known as N'ha-a-itk.

The first non-native to clearly see Ogopogo was Susan Alison in 1872. She was looking across the lake for her husband, who was due to return from a trip. She saw a strange animal swimming against the waves. She had studied native folklore and realised that this must be the dreaded N'ha-a-itk. She was filled with foreboding that the creature had destroyed her husband's vessel and eaten him. Upon her husband's return, she told him of her sighting, but was not believed. She wrote an atmospheric little poem about her sighting that runs thus.

> *Miles to westward lies an island,*
> *An island all men dread,*
> *A rocky barren island,*
> *Where a monster makes his bed.*
>
> *So busy are the fishers,*
> *That they hardly spare a glance,*
> *To the black line of white crested waves,*
> *That so rapidly advance.*
> *From the westward, from the island,*

> *The island all men dread,*
> *From the rocky barren island,*
> *Where the monster makes his bed.*

The island in question is Rattlesnake Island off Squally Point. Mrs Alison's husband may have laughed but soon others had seen the serpent. John McDouall, for example, would certainly not have laughed. He was a trader who would always cross the lake by canoe with his horses in tow. On one occasion in 1860, he was horrified to see his horses dragged one by one beneath the surface. He had to cut the tow rope to prevent his canoe being pulled under.

Early settlers took the threat of N'ha-a-itk very seriously and carried guns whenever they were close to the lake. In 1926, a ferry operating on the lake was armed in case of attack by the monster.

It was also in 1926 that the monster got its modern name Ogopogo. The name comes from an English music hall ballad of 1924, written by Cumberland Clark, and sung by Mark Strong. It concerns the hunt for a banjo-playing monster called Ogopogo in the hills of Hindustan and runs...

> *I'm looking for the Ogopogo*
> *The funny little Ogopogo*
> *His mother was an earwig*
> *His father was a whale*
> *I'm going to put a little bit of salt on his tail*
> *I want to find the Ogopogo*
> *While he's playing on his old Banjo*
> *The Lord Mayor of London*
> *Wants to put him in the Lord Mayor's show.*

Quite why the name was applied to the Okanagan monster is anyone's guess, but it stuck and the lake monster has become far better known around the world than the almost forgotten music hall song.

Ogopogo follows the typical North American lake monster template, with its horse-like head and elongated body, with a crest or mane running along it. Sometimes horns are reported on the head.

An early twentieth-century sighting was by Edythe March.

I had been told of Ogopogo by my father, Henry J Blurton, who worked as a game warden when I was a child. The Indians advised him never to go canoeing on the lake without first tying up a grouse or piece of venison behind the canoe as a precaution against Ogopogo upsetting it, for they believed that should an upset occur the elusive monster would go after the bait rather than the man.

Our teacher's father, who often went fishing at the northern end of the lake, failed to return one day. They found his boat on the lake right side up with all the fishing gear intact, but no Mr Homuth. I truly believed Ogopogo had gotten him.

The first time I saw Ogopogo he was amusing himself by swimming at a fantastic rate, stopping suddenly, then repeating this performance by going in the opposite direction. This happened in the late autumn of 1933. This time there were two of us travelling from Princeton to Vernon. I spotted a wake behind an object that was moving at a tremendous speed, so we stopped to look. It was certainly not a boat, as it resembled something like a huge snake's head. We knew it was Ogopogo.

In the summer of 1936, Geffory Tozer and Andy Aikman were fishing on Lake Okanagan. Their attention was grabbed by a seagull squawking loudly about one hundred and fifty feet away. The bird rose into the air, but a huge grey animal lunged up, grabbed it, and dove back into the lake, as the friends looked on in shock. The bird had been twelve to fifteen feet in the air when the monster had struck. Its girth was about eighteen inches, and it was silvery-grey in colour.

On July 18, 1949, three youths working at an orchard saw Ogopogo showing dark-green, sinuous coils. They shot at it with a .22-calibre rifle but it had no effect. The monster was spotted three times in a half hour period.

One good thing did come out of this encounter. The attorney general of British Columbia said that the monster should be protected under Section 26 of the Fisheries Act that prevents aquatic animals being hunted with guns or explosives.

Mrs E. A. Campbell was sitting with two friends on the lawn of her home on the afternoon of July 6, 1952 when they saw Ogopogo a few hundred feet out on the lake. Mrs Campbell said:

I am a stranger here. I did not even know such things existed. But I saw it so plainly. A head like a cow or a horse that reared right up out of the water. It was a wonderful sight. The coils glistened like two huge wheels going around and around. The edges were all ragged like that of a saw. It was so beautiful with the sun shining on it. It was all so very clear, so extraordinary, as it came up three times, submerged and disappeared.

Birds were on the menu again in 1969 when Roy P MacLean, publisher of the *Kelowna Daily Courier*, saw Ogopogo (in water only a metre deep), some fifteen metres from his house. He had seen the flock of ducks he fed daily fly up in a panic. He then observed, something as thick as a car tyre, showing three humps, undulating through the shallows. It appeared to be feeding before heading back out to deeper water.

Graham Merrick must be one of the few people in the world who has been lucky enough to spot a lake monster from a plane. On March 21, 1990, he was flying out of Penticton on an Air BC flight at 9.30 a.m. Looking out of the right side of the plane, he saw three large animals slither over a sandbar at the south end of the lake. The creatures ranged from thirty to seventy feet long. They were light brown in colour and stood out against the clear water.

Due to mechanical failure the plane had to return to Penticton. Graham saw one of the smaller specimens again as they flew back. Its body looked like an upturned canoe.

There have been multiple witness sightings of Ogopogo. One of the best occurred on July 13, 1994 when Darlene Viala, two other adults, and five children came too close for comfort whilst boating on the lake. A snake-like monster showing between eight and ten coils of its fifteen- to eighteen-metre body surfaced next to their craft. The monster came within twenty feet of the boat, causing the witnesses to become hysterical with fear. It had greenish skin and was about three feet thick.

We have only seen the tip of the iceberg here. There are many, many more monster-haunted lakes both in North America and beyond, but space restrictions mean we can only cover a few in this book.

What are lake monsters? Certainly, some of them are misinterpretations of natural phenomena such as wave formations and debris being dragged along by undercurrents. Others are known animals such as seals swimming upriver and into freshwater lakes. Huge fish are probably involved. I was called in to investigate a 'monster' reported at Martin Mere wildfowl reserve in Lancashire. Witnesses had seen something grabbing swans and other water birds and dragging them underwater. Others had seen something the length of a car swimming in the small, shallow lake. At first, I dismissed the idea of a large predator living in such a small lake. I was proved wrong, however. Within an hour of arriving at the lake I saw the monster for myself. It was an eight-foot-long wels catfish (*Silurus glanis*). This huge fish can reach sixteen feet long and is found in the rivers and lakes of eastern and southern Europe. It was introduced to Britain in the nineteenth century by the Acclimatization Society, a Victorian outfit that attempted to introduce foreign species into the UK.

The beluga sturgeon (*Huso huso*) is the largest freshwater fish in the world. The largest recorded specimen was twenty-three feet six inches long and was caught in 1827 in the Volga estuary. Seeing such a fish in a lake would surely convince the witness they had seen a monster.

Common eels live in freshwater but swim out to sea to spawn. Then they die, but their hatchings swim back to fresh water and the cycle begins again. One theory is that some individuals do not sexually mature and hence never leave their freshwater homes. Known as eunuch eels, nobody knows how long they live or how big they get. In 2004, a family of Canadian tourists said they saw an eel twenty-four feet long in Loch Ness.

Yet we must ask ourselves the question: if a known animal can swim miles upriver into a lake, what is stopping an unknown animal doing just the same? And it is down into the saltwater depths of sea we must go in the next chapter.

CHAPTER FOUR

Sea Monsters

'The dragon green, the luminous, the
dark, the serpent haunted sea.'
**—James Elroy Flecker (1884–1915), 'The
Gates of Damascus'**

•

We have already seen how the world's oceans have given us the legendary
kraken, a prehistoric fish and a huge, weird, plankton-feeding shark. As
recently as 1998, scientists discovered ten-foot-long carnivorous worms
and six-inch long isopod crustaceans resembling a giant woodlouse. These
were part of a community of giant invertebrates living in the Antarctic
seas. What else is down there?

Bernard Heuvelmans, the 'father of cryptozoology', made a mammoth
study of sea serpent reports from 1639 to 1964. He published his work
in a book, *Le Grand Serpent-de-Mer* in 1965. This was later translated
into English, under the evocative title *In the Wake of the Sea Serpents*.

Heuvelmans classified nine different types of sea monster.

1. The Long-Necked: A long-necked beast with a small head, a large
   body and four flippers. It undulates vertically—that is, up and
   down as opposed to side to side. Heuvelmans believed this to
   be a pinniped—a member of the seal and sea lion family—with a
   vastly elongated neck.

2. The Merhorse: An animal with very large eyes, a horse-like face,
   whiskers, and long hair. The merhorse is also a vertical undulator.
   This too, he thought of as pinniped of some kind.

3. The Many-Humped: An animal with a fairly short neck, a long
   body and either a row of humps on the back or a flexible spine
   that loops up in coils out of the water. It undulates vertically.
   Heuvelmans thought that the many-humped was a relative of an
   archaic group of whales called Basilosaurs.

93

4. The Many-Finned: A weird beast with a scalloped tail, plated skin, and spines protruding from its sides. It undulates in the vertical plane. This too, he contended was a primitive whale with armour plating.

5. The Super-Otter: A huge creature of northern latitudes that bore a passing resemblance to an outsized otter. It undulates vertically. This, the professor said, was likely to be an even more primitive form of whale. One that still had legs.

6. The Super-Eel: A fish—a titanic eel larger than any known to science—that undulates horizontally.

7. The Marine Saurian: A reptilian monster resembling a giant crocodile, but far larger than any known species. It undulates horizontally.

8. The Father of All the Turtles: A turtle that dwarfs all known species either extant or from the fossil record.

9. The Yellow Belly: A huge tadpole-shaped animal. Undulating horizontally, and showing a tail, he thought that it was possibly some form of giant ray.

Heuvelmans was a genius and a man ahead of his time. If it were not for him, cryptozoology as a science would not exist, and I would not be writing this book. I owe the man my career and he is one of my heroes. He has, however, been criticized for these classifications. He seemed biased to the marine mammal explanation of many sea serpents and was known to shoehorn sightings into categories that did not really fit to support his ideas.

Heuvelmans also based some of his marine mammal ideas on knowledge that we now know to be mistaken. For example, it was once thought that the basilosaurs had highly flexible backbones that could be arched up to form coils. This gave rise to the theory that the 'many-humped' sea serpent was a basilosaur. We now know that these ancient whales had ridged spines, no more flexible than those of modern whales, and a cucumber-shaped body.

It was also once thought basilosaurs had armour-plated skin. This was due to scutes found in association with basilosaur fossils. These bony plates are now known to have come from fossil turtles. Ergo, the theory that the 'many-finned' sea serpent was also a basilosaur, has been shot down in flames.

The pinniped idea does not stand up well. All seals, sea lions, and walruses need to haul up on land to give birth to their young. Colonies of such massive creatures doing so would surely have been discovered by now. Sea serpents it would seem give birth to live young in the sea. The merhorse may well be a mammal of some description, but it doesn't seem to be a pinniped.

Most sea serpent sightings fall into two categories. The first is a long-necked creature with a thick body and four flippers. The second is an elongate creature that throws its body into a series of vertical humps or loops.

The Many-Humped Sea Serpent

The many-humped sea serpent has been reported in seas all over the globe. In Europe, it is the cold waters of Scandinavia that have had the most recorded sea serpent sightings. The tradition of the many-humped sea serpent goes back centuries in these seas. In 1555, Olaus Magnus, the exiled Archbishop of Uppsala, then residing in Rome, published his book *Historia de Gentibus Serpentrionabilus*. This dealt with sea monsters in Scandinavian waters. The most infamous was of monstrous size.

> *They who in Works of Navigation, on the coasts of Norway, employ themselves in fishing or merchandise, do all agree on this strange story. That there is a serpent there, which is of vast magnitude, namely two hundred feet long and moreover twenty feet thick: and is wont to live in rocks and caves towards the sea coast about Bergen, which will go alone from his holes in a clear night, in summer, and devour calves, lambs, and hog, or else he goes into the sea to feed on polyps, locusts, and all sorts of sea crabs. He hath commonly hair hanging from his neck a cubit long, and sharp scales, and is black, and he hath flaming shining eyes. This snake disquiets the shippers, and he puts his head on high like a pillar, and casteth away men, and he devours them: and this happeneth not, but it signifies some wonderful change*

of the Kingdom near at hand: namely the princes shall die, or be banished: or some tumultuous wars shall presently follow.

Nearly two hundred years later, another cleric took up Magnus's reins and continued his work. Eric Pontoppidan, Bishop of Bergen, studied and collected stories of encounters with these creatures, mainly by fishermen and sailors. The Bishop devoted a whole chapter of his book *Natural History of Norway* to these monsters, and asked the question: are they man-eaters?

I return again to the most interesting inquiry concerning them, which is whether they do mankind any injury? And in what manner they may hurt the human species. Arndt Bernsen, in his account of the fertility of Denmark and Norway, p.308, affirms that they do; and says that the Sea snake, as well as the Trold whale, often sinks both men and boats. I have not heard any accounts of such an accident hereabouts, that might be depended on; but the North traders inform me of what has frequently happened to them, namely that a Sea snake has raised itself up, and thrown itself across a boat, and sometimes even across a vessel of some hundred tons burden, and by its weight has sunk it down to the bottom. One of the aforesaid North traders, who says he has been near enough to some of these Sea snakes (alive) to feel their smooth skin, informs me that sometimes they will raise up their frightful heads, and snap a man out of a boat without hurting the rest...

It is said that they will sometimes fling themselves in a wide circle around a boat, so that the men are surrounded on all sides. This Snake, I observed before, generally appears on the water in folds or coils; and the fishermen, in a known custom in that case, never row towards the openings, or those places were the body is not seen, but is concealed under the water, if they did that the snake would raise itself up and upset the boat. On the contrary, they row full against the highest part that is visible, which makes the snake immediately dive; and they are released from their fears.

The Trold whale seems to be a local name for the giant squid (*Architeuthis dux*). At one time this beast was every bit as legendary as the sea serpent—that is, until it turned up alive and kicking off the coast of Newfoundland in the 1870s.

A dramatic encounter with a sea serpent is recounted in this letter here from Pontoppidan's tome.

Sir, in the latter end of August, in the year 1746, as I was on a voyage, on my return from Trondhjem, a very calm and hot day, having a mind to put in at Molde, it happened that when we arrived with my vessel within a mile of the aforesaid Molde, being a place called Jule-Naess, as I was reading a book, I heard a kind of murmuring voice from amongst the men at the oars, who were eight in number, and observed that the man at the helm kept off from shore. Upon this I inquired what was the matter, and was informed that there was a sea serpent ahead of us. I then ordered the helmsman to keep to land again and to come up with this creature of which I had heard so many stories. Though the fellows were under some apprehensions, they were obliged to obey my orders.

In the meantime, the sea snake passed us by, and we were obliged to turn the vessel to get nearer to it. As the snake swam faster than we could row, I took my gun, which was loaded with small shot, and fired at it; on this it immediately plunged under water. We rowed to the place it sank down (which was calm and might easily be observed) and lay upon our oars, thinking it would come up again to the surface; however, it did not. Where the snake plunged down, the water appeared thick and red; perhaps the small shot might have wounded it, the distance being very little.

The head of this sea serpent, which is held more than two feet above the surface of the water, resembled that of a horse. It was a greyish colour and the mouth was quite black, and very large. It had large black eyes, and a long white mane, which hung down over the surface of the water. Besides the head and neck, we saw seven or eight folds, or coils, of this snake, which were very thick, and as far as we could guess there was a fathom's distance between each fold.

I related this affair in a certain company, and there was a person of distinction present, who desired that I would communicate to him the authentic detail of all that had happened, and for this reason two of my sailors who were present at the time and place where I saw this monster, namely NIELS PETERSON KOPPER and NEILS NEILSEN ANGLWIGEN, will appear in court to declare on oath the truth of every particular herein set forth and I desire the favour of an attested copy of the said description.

I remain, Sir, your obliged servant—

L. VON FERRY

Bergen. 21st February 1751

It is unlikely that small shot did much damage to such a creature. Mr Ferry and co. were lucky it did not turn on their boat. Such an occurrence happened in 1815 in Ronsdal Fjord. A small sail boat with five men aboard encountered a sea serpent. One man, J. C. Lund, shot at the monster's head from close range. The beast seemed unhurt, but angry, and pursued the boat until it reached shallow water.

In Norway, at the mouth of the Oslofjord, eleven people aboard the yacht *Tommy* encountered a sea serpent. The Reverend Hans Davidsen describes what they saw on the 4th of August 1902.

We soon saw that it was an unknown sea animal moving at—so far as we could judge—about four miles (probably four sjomil, i.e. sixteen miles) an hour. It was one or two cables away from us.

From time to time three big humps showed on the surface, and three of us also saw the creature's head, oblong in shape, and as far as we reckoned, about three feet long. The humps formed a continuous series and were dark in colour, with a shining surface. They seemed to be at least two feet in diameter. Seen from the side the animal's motion seemed to be undulating. It is impossible to give an exact estimation of the creature's length. From what we saw the head and the three visible humps were certainly twenty feet long altogether. From the distance between the head and the humps, and the length and thickness of the latter, the total length must have been 60 feet. We all saw that the humps were joined, and could not belong to a series of creatures swimming in line.

Because of its great speed, the animal left a broad wake behind it. We did not see foam, but we noticed that the front part of the body raised a considerable wave. The head was held near the surface in a slightly oblique position. One of the passengers thought he saw a fin on the creature's back. We watched for five to ten minutes, with the naked eye, and through powerful binoculars.

W. E. Parkin worked in the office of an iron-ore mine at Bogen in the Norwegian district of Norland from 1910 to 1914. In June or July of 1914, a lady burst into the office, telling him to come and see an extraordinary thing that had just swum into the bay. He ran outside.

What met my gaze was an object sticking out of the water at an angle of approximately forty-five degrees. It appeared, from where I was, to be five or six feet out of the water. Behind it was a gap, then several regular humps. The largest number I counted at one time was seven, the smallest five.

The animal swam slowly round the bay before the astonished eyes of several ladies fishing on the pier, several employees of the mining company, the peirmaster, the postmaster, and some old men and children. Several people launched rowing boats to try and see the animal from up close, but this seemed only to frighten it off, and it gradually made off to sea.

Asia is a hotspot for the many-humped sea serpent. On August 18, 1901, First Officer F Wolfe in charge of the Chinese customs launch Lung-tsing was off Tai Yue Shan Island, Hong Kong. He spotted a dragon-like animal coiled on the sea's surface. It held its head about three feet above the water.

It bore a crest on its head, and two fins high on its neck. He ordered his second officer, V Kuster, into a gig with a number of sailors, and (stupidly) commanded them to attempt to kill it with a boathook. This seems to me akin to attempting to knock over Nelson's Column with a fly-whisk. In any event, the serpent bit at one of the oars and reared up fifteen feet out of the water before diving and vanishing. The men estimated its length at twelve to fifteen metres.

Another series of sightings had occurred in Along Bay, on the coast of Vietnam four years previously. The French gunboat *Avalanche* encountered dragon-like serpents several times in this island-dotted bay on the coast of Tongking. The first was July 1897. Lieutenant Lagresill takes up the story.

> *In the month of July last (1897) the Avalanche saw for the first time, off Along Bay, two animals of weird shape and large dimensions; their length was reckoned at about sixty-five feet, and their diameter six to ten feet. The feature of these animals was that their body was not rigid like that of known cetaceans, but made undulatory movements similar to a snake's, but in a vertical direction. A revolving gun was loaded and fired at six hundred yards, at slightly too short a range. They immediately dived, breathing loudly and leaving a wash on the surface like breakers. They did not reappear, but we thought we saw their heads, which we judged to be of small dimensions.*

> *On February 12th of this year (1898), when crossing the Bay of Fai-tsi-long, I saw similar animals again. At once I gave chase and had the revolving guns loaded. Several shots were fired at one of them, at ranges of between three hundred and four hundred yards, and the last two shots reached them without seeming to do them the least harm, the shells bursting on the surface. I also tried to reach them with the bow of the ship, but their speed was greater than that of the* Avalanche. *Each time, however, that this animal came into shallow water, it turned back, which enabled me to gain upon it and confirm its great size. It frequently emerged, and always one noticed its undulatory movements. Each emergence*

was preceded by a jet of water, or rather of water vapour made by a loud breath, unlike ordinary blowers which inhale water, and blow it out to a certain height.

The colour of the animal was grey with several black fins. Its trail was easily followed by the release of its breath, which formed circles four to five yards in diameter on the surface of the sea, which was then perfectly calm. At one moment I thought I had reached it. The chase went on for an hour and a half and had to be abandoned as night was falling.

Lagresille is quite wrong in asserting that blowers (whales) take in water and blow it out. A whale's spout is actually air (and a little water) that collects in the concave blowhole which is 'spouted out' as the whale surfaces to exhale. He was invited, a week after his latter adventure, to a reception organised by Admiral de la Bedolliere, given in the honour of Paul Doumer, Governor General of Indo-China and later president of France. On hearing his story, many of his fellow officers scoffed, but Lagresille was vindicated the following day.

He invited several of the naysayers to visit the Fai-tsi-long archipelago on the *Avalanche*. Whilst at lunch it was reported that the two sea dragons had returned. They all rushed on deck, and there before the sceptics' eyes, the monsters swam some six hundred feet away. Lagresille said:

We gave chase to one of them for thirty-five minutes, and at one particular moment we saw it clearly two hundred yards off the beam, floating horizontally. It had three undulations without a break, which ended with the appearance of its head, which much resembled a seal's, but almost double the size. We could not see if it had a neck, joining it to the body, of relatively much greater dimensions: this was the only time we saw the undulations appear without a break. Until then we might have thought that what we took them for were humps appearing in succession: but from the testimony of all the witnesses, doubt is no longer permissible, for before they appeared, we saw the animal emerging by the same amount along its length. Two of the officers present possessed a camera: they ought to have been able to use it then, but they were so surprised by what they saw, that when they thought of aiming their cameras the animal dived, only to appear much farther away in much less clear conditions unfavourable to taking a photograph.

To sum up, the animals seen by the 'Avalanche' are not known. Their length is about sixty-five feet (minimum), their colour is grey and black, their head resembles that of a seal, and their body is subject to undulations that are sometimes very marked: finally, their back is covered with a sort of saw-teeth which removes any resemblance to known cetaceans; like the latter they reveal their presence by blowing noisily, but they do not spout a jet of inhaled water like whales; it is rather their violent respiration which causes a sort of vapourization of water that is ejected in drops and not a jet. Undoubtedly these animals, known and feared by the Annamites, must have provided the idea of the Dragon, which modified and amplified by legend, has been, if I may so term it, heralded into the national emblem.

He wrote off to Admiral de la Bedolliere telling him what he had seen. The Admiral wrote at once to Lagresille and apologized for doubting his word. He wanted a concerted effort to capture a specimen of the species with the gunboats. He planned to chase one into shallow water where it would be stranded and could be caught. A diplomatic crisis in China saw to it that the plan was never carried out.

The Charles-Hardouin was on passage from Nantes to Hong Kong in November and December of 1903 in Tourane Bay. The helmsman alerted the mate to a dark mass very close to the ship:

Fifteen to twenty yards from the ship a double mass appeared, the length of each part must have been about twenty-five feet and the distance between them eighteen. The bulk of each of these coils could be compared to that of a big half hogshead barrel: a spiky crest gave the coils a quite singular appearance.

It all undulated like a snake in motion, and its speed was markedly greater than that of the ship, which was then doing nine knots as I recall. The colour was dirty black. A few seconds later the animal dived, churning the waters violently.

What I heard during a long period on the south China coast leads me to believe that they are amphibious (i.e. air-breathing), and that their appearances were once fairly frequent on the coasts of the China sea, and I think that the Taiwan embroidered beast on Imperial and Annamite flags is just a stylised version of this animal.

Lieutenant L'Eost made an official report to Rear Admiral de Jonquieres of a monster he saw from the gunboat La Decidee in May of 1904.

Sir, on the afternoon of 25th of February last, when steaming out of Along Bay, La Décidée met near the Noix Rock a strange animal apparently of the same species as seen in the same locality in 1897 and 1898 by Lieutenant Lagresille on board the Avalanche, *which observations were published in the* Bulletin de la Société Zoologique de France *(1902), of which I had no knowledge until after I made my own.*

I first saw the back of the animal at about three hundred yards, on the port bow, in the shape of a rounded blackish mass, which I first took to be a rock, then seeing it move, for a huge turtle twelve to sixteen feet in diameter.

Shortly afterwards I saw this mass lengthen, and there emerged in succession, in a series of vertical undulations, all the parts of the body of an animal having the appearance of a flattened snake, which I reckoned to be about a hundred feet long and the greatest diameter twelve to sixteen feet.

The animal dived. I did not observe it again, my attention being distracted by handling the ship. The observations which follow were gathered from various members of the staff and crew.

The animal appeared a second time about one hundred and fifty yards away and dived beneath the ship just aft of the gangway. Its back, on this second appearance, was all that was visible at first. It was semi-circular in section, not at all like that of the cetaceans (Dr Lowitz). Its skin was black showing patches of mottled yellow (Able Seaman Sourimant); according to Seaman Leguen, it was dark yellow and quite smooth.

The back then disappeared, leaving big ripples, and the head alone appeared near the gangway. Here are the observations of Leading Engineer Ponaurd, who was in this position. All other seamen present have confirmed every detail.

When he heard voices on the bridge, he looked out and saw waves like the sea breaking or a rock awash, or like those made by a submarine diving. He turned to call his mates, and they all came and watched.

The head and neck came out of the water only forty yards away. The head was the colour of the rocks in the bay (greyish, white mixed with yellow). It was like a turtle's; the skin seemed rough, and this roughness seemed due to scales rather than hair.

The witnesses estimate the diameter of the head varied between fifteen and thirty inches; it was slightly greater than that of the neck.

The head blew two jets of water vapour. The rest of the body appeared fleur d'eau. It undulated in a horizontal direction. The animal moved at a speed estimated at eight knots.

When it was almost alongside, the head dived, and a series of vertical undulations were seen running along the body, just out of the water.

The animal once reappeared near the ship's starboard quarter. Marine Lecoublet and Seaman Le Gall were there.

The body moved forwards in vertical undulations. In its whole length there were five or six marked undulations. This length is estimated by these two witnesses at more than thirty feet. They describe a head wider at the back than the front and longer than that of a seal.

The body seemed to them to be of almost the same dimensions all along its length. They compare it to a Blower's. This estimate, together with what seems to me much too small a figure for the length, makes me think these witnesses saw only part of the body.

The skin was smooth. Nobody saw fins. The animal did not blow at this time. It dived again and appeared some way astern. One could now only make out a long, blackish body, with moving curves and jets of water vapour.

From what the witnesses at the gangway saw, the animal breathes through its nostrils rather than through the top of its head. Nobody observed the head in detail.

This is an interesting case as it has many witnesses, all experienced seamen, who observed the monster from a number of differing vantage points. It also shows that the many- humped is capable of undulating in both the vertical and the horizontal plane.

Australian waters play host to this kind of beast too.

In 1934, the following report of a scaled, many-humped sea serpent was published in *The Victorian Naturalist* by Mr. A. H. E. Mattingly. The sighting was made by Oscar Swanson of Townsville, Queensland, his son Harold, and William Quinn. On August 18, the three went to fish. Soon after leaving Townville breakwater in their motor launch, they saw four dark objects in the water and moved closer to investigate. They got within about four hundred and fifty feet of the animal when it submerged.

Then we thought it would come at us, and we turned to make for the Beacon, which has a ladder to the top on which the lamp is lit.

We were wishing we were in a speed boat. We stowed the little fellow up forwards under what bit of decking we had, and hoped for the best. I might mention that the sea at this time was as smooth as glass. After about five minutes the monster arose again in the same place (coming up just like a submarine). We were about three-quarters of a mile past the Beacon; on reaching it we caught hold of the ladder and watched to see what movements the monster would make. After waiting half an hour and seeing no movements, excepting the head swaying side to side as if watching us, we decided to make back to town, get rid of the boy, and get a camera, as it looked as though the monster would stop there all day. On reaching the jetty wharf, I rang Mr Jim Gibbard, sub-editor of the Townsville Bulletin, who picked up press photographer Mr Ellis, and armed with two cameras, we once more set out (without the boy).

Sadly, the monster seemed to have been disturbed by a ship and had moved farther out. The watchers saw two large humps, but it was out of camera range. Mr Swanson described the animal and made a drawing of it.

You will see by the rough sketch submitted what the monster was like. The head rose eight feet out of the water and resembled a huge turtle's head; the mouth remaining closed. The head was about eight feet from the back of the head to the front of the mouth, and the neck was arched. The colour was greyish-green. The eye (we could only see one, being side on) was small in comparison to the rest of the monster. The other part in view was three curved humps about twenty feet apart, and each one rose from six feet at the front to a little less at the rear. They were covered with huge scales about the size of saucers, and also covered in barnacles. We could not get a glimpse of the tail, as it was under water.

In October 1939, Cecil W. Walters was on anti-submarine watch on the naval oil tanker *HMAFA Karumba*. The ship was two days out of Darwin and northwest of King Sound, Western Australia. At 1 p.m. he and Jack Mack, another seaman, were manning telescopes on either side of the stern-gun. They saw a huge animal about four miles distant of the ship and travelling at about fifteen knots. It threw up a huge bow-wave and reared its head and neck around three metres out of the water. The body was thrown into two loops through which water was visible. The telescopes had been calibrated for distance, and Mr. Walters thought the visible parts of the monster must have been over ninety feet. The monster was brownish-yellow in colour, with blotches of pale blue, green, and yellow. The animal's jaw seemed to be moving, and a tongue flickered in and out. Mr. Walters took a photograph through the telescope and, after the war, sent it to his brother in law...who lost it!

The warm waters off the southern Pacific coast of the USA have been visited by the many-humped sea serpent as well. Such a creature first reared its head in 1976 near San Francisco. The Great Western Pacific Report ran a story about Tom D'Onofrio, a minister from Bolinas, California. Tom's account runs thus:

> On September 30, 1976, at noon I experienced the most overwhelming event in my life. I was working on a carved dragon to use as a base for a table, and couldn't complete the head. I felt compelled to go down to Agate Beach where I met a friend, Dick Borgstrom.
>
> Suddenly, one hundred and fifty feet from shore, gambolling in an incoming wave, was this huge dragon, possibly sixty feet long and fifteen feet wide.
>
> The serpent seemed to be playing in the waves, thrashing its tail. We were so overpowered by the sight we were rooted to the spot for about ten minutes. I literally felt as if I were in the presence of God. My life has changed since.

A colourful account, and one that would have probably attracted little attention if the monster had not returned. Upon its second visit, the dragon was seen by an entire road construction crew from the Californian Department of Transportation. On November 1, 1983 they were on a cliff top road, part of Highway 1, just south of Stinson Beach. It was 2:30 in the afternoon. Safety engineer Marlene Martin takes up the story:

> *The flagman at the north end of the job site hollered, 'What's that in the water?'*
>
> *We all looked out to sea, but could see nothing, so the flagman, Matt Ratto, got his binoculars. Finally, I saw the wake and I said, 'Oh my God, it's coming right at us, real fast.'*
>
> *There was a large wake on the surface and the creature was submerged about a foot under the water. At the base of the cliff it lay motionless for about five seconds and we could look directly down and see it stretched out. I decided it must have been one hundred feet long, and like a big black hose about five feet in diameter. I didn't see the end of the tail.*
>
> *It then made a U-turn and raced back, like a torpedo, out to sea. All of a sudden, it thrust it's head out of the water, its mouth went towards the sky, and it thrashed about.*
>
> *Then it stopped, coiled itself up into three humps of the body and started to whip about like an uncontrolled hosepipe. It did not swim sideways like a snake, but up and down.*
>
> *I had binoculars and kept them focused on the head. It had the appearance of a snake-like dinosaur, making coils and throwing its head about, splashing and opening its mouth. The teeth were peg-like and even, there were no fangs. The head resembled the way people drew dragons except it wasn't so long. It looked gigantic and ferocious.*
>
> *I did not see any fins or flippers and it had bothered me that it could move so fast in that way. It was scientifically impossible for anything to go that fast without them. It was not like a snake going sideways; it went up and down.*

It stunned me. Never in my life, could I ever have imagined a thing so huge could go so fast. I thought, when I saw it, this is a myth.

There were six of us at this time, all looking over the rail in disbelief. I was so glad everybody saw the same thing.

I've never really told anybody this before, and I cannot swear to it but the eye I saw looked like it was red, a deep burgundy-ruby colour. When I think about the thing, I still see that colour and what's amazing about it is that I've never seen that particular red on anything before.

Another member of the crew, truck driver Steve Bjora, estimated that the monster was moving at fifty miles per hour. The crew at first said nothing of their sighting, but they were overheard talking about it at a pizza parlour and the news got out. As a result, other witnesses came forwards. Marlene visited Tom D'Onofrio and saw his carved dragon. She said it was similar to the animal she had seen.

The most detailed account of the monster was given by Bill and Bob Clark who saw it on February 5, 1985.

From the start this particular morning was different. The day before had been beautiful with no wind and temperatures of seventy degrees. The fifth was just gorgeous, with a clear sky, calm water, and high tide. We had never seen the San Francisco Bay so calm. It was like looking in a mirror. Anything sticking above the surface of the water was easily seen. As a result, at around 7.45 a.m., we noticed a group of sea lions about one hundred and fifty yards in front of us. While watching them, we thought we saw another sea lion come around Stone Tower point and approach the group. When it got within a few yards, a long, black, tubular object telescoped about ten feet straight out of the water and lunged forwards almost falling on top of the sea lions. They immediately began swimming away, leaping in and out of the water as they fled towards shore.

The creature churned the water as it swam behind them, moving so fast it was a blur, but we could see three or four vertical undulations moving down the length of the animal. Suddenly, it went underwater. Meanwhile the sea lions were coming closer and closer to where we were parked along the Marina Green only yards from the bay. They came so close that Bob was able to make eye contact with one and see the fear of death in its eyes as it leapt out of the water. The creature followed close behind, stirring up the water as it made a final attempt to procure a meal.

Now only twenty-five yards away, an arch of the animal was exposed, which looked like half a truck tyre. It appeared black and slimy, yet at the same time glistened in the early sunlight. The creature was swimming slightly below the surface almost parallel to the shore. The water was very clear, allowing the outline of the serpent's head to be observable. A short flat snout, eyebrow ridges, and lots of neck could be seen. It must have been thirty feet of neck because we both thought a big snake had just swum by. We were expecting to see the end of the snake but instead of getting smaller it began to get much larger. What we watched wasn't a big snake, but something even more unbelievable.

There was a loud crash and with a spray of water the creature seemed to stop dead in its tracks (later at low tide the next day we realised that a ledge with large rocks on it extended twenty yards into the bay at the location where the creature crashed). Instantaneously, a long black neck popped up, twisting backwards away from the shore, then splashed as it hit the surface of the water and disappeared. The serpent twisted clockwise like a corkscrew, and exposed its midsection above the water, giving us an excellent view of the underbelly, which was creamy white with a tint of yellow. It resembled an alligator's belly with a soft leathery look but was divided into many sections several feet wide. The midsection was about twenty feet long, black on top, and slowly changed from a mossy green to a grassy green and ultimately to a yellow-green as it approached the underbelly. It had hexagonal scales next to each other rather than overlapping. The largest scales appeared at the widest part of the midsection where the underbelly and the side of the creature met, gradually reducing in size as they approached the top, front and end of the midsection. The largest scales were bigger than a silver dollar and the smallest were the size of a dime. There was a distinct line where the texture of the skin changed from scales to the smooth, leathery underbelly.

While it continued twisting, another section six to nine feet long arched upwards three feet above the water. The arch twisted away from us, exposing a fan-like appendage that was attached to its side at the waterline. It looked like a flag flapping in the wind. It was triangular in shape with a serrated outer edge. Mossy green ribbing ran from a single point attached to the side of the animal like the spokes of a wheel. A paper-thin green membrane stretched between each rib which extended farther than the membrane, creating a serrated edge. The appendage was equilateral with each side, almost two feet in length, reminding us of a dragon's wing. Bob concentrated on counting the ribs but stopped when he got to six as there were too many. Bill looked at the rest of the animal and saw two appendages, one at the beginning and one at the end of the midsection. They looked like stabilizer fins as opposed to flippers for propulsion. Slowly the body sank beneath the water onto the rocks below. Under the surface of the water we could see the upper section of the neck. Four tightly folded coils were formed directly behind the head.

The creature moved its neck with a whipping motion, and the four coils travelled backwards in a packet, dissipating upon reaching the midsection. Instantly it created another packet of four coils behind the head, and again these were whipped backwards towards the midsection. This was repeated several times, until the creature began to pull itself into deeper water. It was like watching a freight train pull out of a station; each section had to wait for the section in front to move.

The outline of the head could be seen as it sat underwater but no details were observable except a snake-like head with large jowls. When it began to swim north towards the middle of the bay, we thought we saw a ridge line along the top of the rear section. However, we never saw the tail. As it swam away at a leisurely pace, several arches could be seen undulating above the water. A few seconds later it slopped beneath the water. Since we never saw the rear end of the animal, it is hard to estimate the total length but it had to be at least sixty feet and probably closer to a hundred feet.

This is perhaps the most detailed account of a sea serpent on record. It is highly interesting for a couple of reasons. Firstly, it is one of the few accounts of a sea serpent observed hunting prey. It seems that this species can deal with large marine mammals as food sources just as a killer whale or white shark would. The observations of how the creature moves are also fascinating. This mode of locomotion is unknown in any other animal, but seems highly efficient, moving the animal at high speeds.

The Clark twins had several more sightings of the monster. They saw it again on February 28 in the same area, but a little farther out. Its head was sticking three feet out of the water and looking towards Alcatraz island. A single arch was observed running down a twenty-foot section of the neck just below the water. Another arch followed, then the head turned left and dove. This makes you wonder about the fate of the prisoners who escaped Alcatraz but were never heard of again!

On March 1, whilst parked at the St. Francis Yacht Club, they saw the neck rise five feet out of the water. An hour later they observed the monster creating a V-shaped wake as it swam, just below the surface, three hundred feet offshore. It submerged when a boat approached.

On their next encounter, on December 22, the brothers were ready with a camera.

It was 8 a.m. on a foggy, misty morning when we had our next sighting. The water was calm. Bob was looking at a buoy fifty yards from the shore, and seventy-five yards west of where we were parked. On its right side he saw what appeared to be two floating telephone poles bobbing in the water. He half jokingly told Bill the serpent was back and pointed to the buoy. When we looked, the two logs had disappeared. In front of the buoy, about a foot above the water, we saw the head looking directly at us. It started to swim very slowly in our direction and without warning the head and neck raised five feet out of the water like a periscope. The head had a flat snout with two black, oval nostrils as large as a man's fist. As it continued to swim towards us Bob grabbed the camera out of the glove compartment and handed it to Bill. When he looked back it was gone. It then reappeared fifty yards in front of us.

Bill attempted to take a picture, while Bob looked through the 7x35 binoculars. He saw the shape of a long-necked archaic marine reptile swim by the slender neck arching gracefully like a swan. Bill jumped out and ran down the beach road trying to get pictures of it. As Bill ran the creature surfaced one hundred and fifty yards away and lifted its neck several feet above the water. Before the creature submerged, he took a picture. It resurfaced one hundred yards away and once again he took a picture of it before it submerged. Bill continued to run and after twenty-five yards the creature reappeared only fifty yards away. He stopped and snapped a third picture as the creature swam slowly with its head slightly above the surface of the water and a single arch undulating westward. Bill continued along the beach to its end, then down some rocks to the water's edge to where the creature was twenty-five yards away with its head still above the water. The creature submerged after another picture was taken.

We decided to have prints made immediately but half the film was unexposed, so we walked around the area taking pictures of anything. We went to a one-hour film developer and asked him to develop every photo regardless of how bad they were. After an hour we returned to get the pictures. When Bill opened the envelope all four photographs of the creature were missing. We ran back to the store and asked the developer about it. He said he didn't develop them because there was nothing on them. Holding back our anger, we told him to make the prints anyway. When we finally got the prints, although overexposed, the creature could be seen in all of them. Because of the rain, fog and time of day, we felt lucky to get anything on film.

Picture 1: The animal is offshore with rocks in the foreground. It was hard to see but it looked like at least ten feet of the neck was above the water. Number 12 on the negative.

Picture 2: Four black spots are above the water. It looked like several arches, and part of the upper neck sticking out of the water. Number 13 on the negative.

Picture 3: The head and an arch slightly above the water. Number 14 on the negative.

Picture 4: Two black spots, it looked like the head and a portion of the neck behind it. Number 15 on the negative.

Bob saw the creature (or another of the same species) again on December 23, 1986. The head and neck of the dragon rose up beside a buoy off Alcatraz but dived before he could get a picture.

A month later Bill was looking towards the Golden Gate Bridge at 8 a.m. when he saw a specimen approaching a buoy. The head and several loops were visible. Bob watched the animal swim around near the buoy, as Bill attempted to take photographs. Two sea lions appeared where one of the monster's coils had been. One seemed to have been wounded, possibly by a bite from the dragon. They dived and did not reappear. Sadly, water on the lens had ruined the film. This is a shame as this account seems to reinforce the idea that the dragon was in the area to hunt and eat sea lions. They had better luck on their next encounter on February 25, 1987.

*Our next sighting occurred in the same area as on January 24,
1987. Once again it was hanging out near the same buoy. It was
only for a short time and not very far out of the water. At first, we
weren't sure if it was the creature, but Bill took two photos anyway.
About half an hour later Bill saw something pop up in front of
us about thirty-five yards away. Bob got out the binoculars and
saw two objects intertwined. One looked like a black hose. It was
twisted round the second object which was an arch of the creature.
Bob saw the first object create an undulation in its body and pull
itself across the arch. Then everything went underwater. Fifteen
minutes later, only twenty yards away, a small black head stuck
a foot or two out of the water and stared at us. After angling its
head slightly, the mouth opened, and it began to 'growl'. Actually,
it was more like a growl and hiss at the same time. We got the
distinct impression it was warning us not to mess with it. Bill took
a picture through the car window, then got out and leaned on the
roof and took two more pictures before it sank straight down like
a rock. The intertwined creatures must have been an adult and a
juvenile creature we happened to observe.*

This time two photos seemed to show the phenomenon. One caught a
head sticking up from the water. Another, part of the long body.

The adult and juvenile creature theory is interesting. We tend to think
of reptiles abandoning their young or eggs but some, like crocodilians,
make excellent parents. The smaller object may well have been a youngster
of the same species, but it may also have been a parasite of some kind like
the remora or sucker fish (*Remora remora*).

The brothers' final sighting was on March 1, 1987, when Bob spotted
two coils moving in unison.

The reaction of the scientific community, even those who should
have known better, was not what it could have been. Professor Bernard
Heuvelmans, the zoologist who created the discipline of cryptozoology
had this to say about the sightings in an August 1985 letter.

*Your report, and your brother's report, are not only the best I
have seen in many years about sea serpents as far as detail and*

precision are concerned, but also the most intriguing. Your sea serpent does not fit at all within any of the categories I have been able to distinguish. However, it looks very much like some of the fabulous sea monsters from ancient reports which I have always thought to be the product of hoaxes or delirious imagination. All the same I am perfectly convinced that you and your brother are quite honest and sincere, and you reported with a wealth of unambiguous details what you saw, or let us say with more scientific vigour, what you think you saw.

In fact, the creature fits nicely into Heuvelmans's category of the many-humped sea serpent. He was obviously distressed by the clearly reptilian nature of the animal. Heuvelmans believed that most sea serpents, including the many-humped, were marine mammals. Here, when he is presented with evidence which suggests otherwise, he is at a loss and tries to sweep the sighting under the carpet with phrases such as 'what you think you saw'.

Forrest Wood, a marine biologist (not a tree surgeon) had the following to say.

...the creature they describe cannot be assigned to any known class of animal: it is a chimera. Taking the descriptions on face value, we can identify it as a vertebrate, but beyond that its features become contradictory. No known fish or aquatic reptile swims with vertical flexations of the body... Cetaceans, sirenians, and true seals do flex their bodies in the vertical plane, but these flexations are modest and they are quite incapable of forming the humps or coils, such as those described.

The scales, slimy in appearance, and pairs of rayed fins could only belong to a bony ray-finned fish (except I am not aware of any fish in which the fins are attached to the body along one edge). With the exception of seahorses and some of their relatives, no fish has what could be called a neck. In all cases, cervical vertebrae are lacking; neck movement is impossible.

*In any case, the described creature cannot be assigned to any class
of vertebrate. On the basis of zoology, including palaeontology
and phylogenetic principles, it is an impossibility.*

As a zoologist myself, I find the above comments as indicative of an
armchair zoologist. Wood is comparing the description with known animals,
whereas the sea serpent is clearly an unknown animal. His statement
that marine reptiles do not flex vertically is only accurate when applied
to modern marine reptiles. Some fossil species seemed to have been quite
capable of vertical flexation. Fish are not the only animals with fins and
scales; reptiles possess them as well. The flexible neck rules out a fish. The
animal can only be one thing: an unknown marine reptile. Mainstream
scientists have minds like trains—they run on rails. They cannot think
outside of the box.

In 1996, the brothers gave the photographs to John Morgan III of
Michigan. Morgan claimed he could computer-enhance them for free.
After months of abusing Bill and Bob's trust, he vanished, taking the
photos with him. Thus, ending one of the most promising cases in
cryptozoological investigation.

The old whaling grounds of New England were once a hotbed of sea
serpent reports. Such a creature was observed by the America Commodore
Edward Preble in 1779 during the War of Independence. Preble was aboard
the ship *USS Protector*.

A huge serpent was spotted and examined through glasses. The
captain ordered Preble to man and arm a boat and attack the creature.
A twelve-oar boat, with a swivel mounted gun in the bow, and an armed
crew approached the animal. It raised its head about ten feet out of the
water, and began to swim away. The gun was discharged, but did not seem
to harm the animal, which soon outpaced the boat. Preble estimated it to
be one hundred and ten feet long, with a head the size of a barrel. Small
wonder the firearms did it no damage!

In the summer of 1817, off Cape Ann near Gloucester, Massachusetts,
a whole wave of sightings occurred. On August 6, two women saw the
creature enter the harbour of Cape Ann. Most people ignored their tale,
but it was seen again by a number of fishermen. Amos Story, a seaman, saw

it near Ten Pound Island. On August 10. Two days later, Solomon Allen, a shipmaster, saw it from a boat. On the 14th, thirty people including the Hon. Lonson Nash, Justice of the Peace for Gloucester, saw it. Mathew Gaffney, a ship's carpenter, fired a musket ball at its head from point blank range. The monster was not in the slightest bit affected.

The Linnaean Society of New England set up an investigation committee in Boston consisting of three carefully chosen members: Jacob Biglow—a doctor; Francis C. Grey—a naturalist; and the Hon. John Davis—a judge. They asked Lonson Nash, a Justice of the Peace, to collect sworn evidence from all the witnesses.

The monster was not idle whilst the learned trio planned their investigation. On the 15th, a merchant named James Mansfield saw it close to shore. The next day, Colonel T. H. Perkins and the entire crew of a ship, including its captain, were treated to a view. On the 17th, three men in a boat said it had come so close to them that it had touched their oars. On the same day another witness watched it from the shore. William Pearson and a friend saw it from their sailing-boat the following day. On the 22nd, a woman watched it through a telescope, as it lay half out of the water on the beach at Ten Pound Island. A Mr and Mrs Mansfield saw it sunning itself on the same occasion. Amos Story saw it again on the 23rd, apparently dozing on the surface. On the 28th, two miles off the eastern point of Cape Ann, Captain Sewell Toppan of the schooner *Laura* and two of his crew saw it whilst they were bound for Boston.

Lonson Nash questioned the witnesses separately and asked them not to discuss their sightings with each other. As not to influence their statements, he always asked them to describe their encounters in their own words, before he asked any questions. He then asked a carefully drawn up series of questions from the Linnaean Society. Of the many affidavits he collected, Matthew Gaffney's was the most dramatic.

I, Matthew Gaffney, of Gloucester, in the County of Essex, ship carpenter, depose and say: That on the fourteenth day of August, AD 1817, between the hours of four and five o'clock in the afternoon, I saw a strange marine animal resembling a serpent, in the harbour, in said Gloucester. I was in a boat and was within thirty feet of him. His head appeared full as large as a four-gallon keg, his body as large as a barrel, and his length that I saw, I should judge to be forty feet at the least. The top of his head was of a dark colour, and the under part of his head appeared nearly white, as did also several feet of his belly, that I saw. I supposed and do believe that the whole of his belly was nearly white. I fired at him when he was nearest to me. I had a good gun and took good aim. I aimed at his head, and I think I must have hit him. He turned towards us immediately after I had fired, and I thought he was coming at us; but he sank down and went directly under our boat and made his appearance about one hundred yards from where he had sunk. He did not turn down like a fish but appeared to settle directly down like a rock. My gun carries a ball of eighteen to the pound; and I suppose there is no person in town, more accustomed to shooting, than I am. I have seen the animal at several other times, but never had so good a view of him, as on this day. His motion was vertical, like a caterpillar.

It is heartening to hear of a scientific organisation taking an active interest in unknown animals. But sadly, the Linnaean Society was about to make a big blunder. They reasoned that the animal must come onto land to lay eggs. In fact, there is no evidence that sea dragons breed on land, and it is probable that they give birth to live young in the water. However, a frantic search for the apocryphal eggs began, but that was not the blunder. Two boys playing in a field near Lollboly Cove found a small black snake. Their father saw it had humps running along its back and promptly killed the poor animal, believing it to be a baby sea serpent. The body was sent to the Society who accepted the claim without batting an eyelid. The specimen was dissected, and detailed plates were drawn up of its anatomy. They even gave it a Latin name: *Scoliophis atlanticus.*

Unfortunately, the zoologist Alexander Lesure identified the specimen as a black racer (*Coluber constrictor*) with spinal deformities. Thus, the promising investigation collapsed under a sea of red faces and mocking laughter. It did not seem to matter that Lesure himself seemed open minded about the existence of the sea serpent, claiming to want to visit the area himself; the Linnaean Society had had quite enough, and stopped the investigation. This is a shame, as the following summer the serpent was back.

On the 29th of July, a boat full of armed men set out to hunt it after several appearances earlier in the month. A specimen received seven or eight rounds without being hurt. The following day, Captain Webber and several whalers approached it, but found their harpoons did no more damage than bullets. The serpent caused such a wash as it swam away that the boat was almost capsized.

On August 12, it had moved to Salem and was seen off the harbour there. By the 16th it was back in Gloucester and seen off Squam Lighthouse. Some whalers pursued it to no avail. Their luck was a little better on the 19th when their quarry surfaced by Squam jetty. Captain Richard Rich hunted it from a large whaleboat and managed to attach a harpoon into its hide. It swam off faster than a whale, taking twenty fathoms of line with it. But the harpoon fell out.

In 1819, it was back again off Massachusetts. Captain Hawkins Wheeler, of the sloop Concord, and his first mate, swore an oath that they had seen it at no more than three hundred feet away.

> *His head was as long as a horse's and was a proper snake's head, there was a degree of flatness, with a slight hollow at the top of the head, his eyes were prominent, and stood out considerably from the surface, resembling the eyes of a toad, and were nearer to the mouth of the animal than to the back of the head.*

The visage is clearly that of a reptile, not a mammal. They reckoned it to be sixty feet long.

For more than a decade afterwards, the monsters were a fixture along the eastern coast of America during the summer months. In 1820, Colonel H. T. Perkins saw it again. Then by four men at Swampscott, three of whom pursued it in a boat, counting twenty humps along its eighteen-metre body.

In 1821, Massachusetts was the scene again, as it was observed by a Captain Bennet. The Colonel and his family saw it again at Nahant. It then turned up off Nantucket and was seen by Frances Joy, a local merchant. A specimen turned up at the harbour of Portsmouth, New Hampshire, and was watched by customs inspector Samuel Duncan and his son.

In 1822, it was seen almost every day in summer off Nahant, and the following year was observed off Plymouth Bay. In 1825, there were no reports from New England, but there were some from Nova Scotia four hundred miles farther north. It was back to its old stamping ground the following summer. A whole ship's crew saw a serpent off Cape Cod.

In 1827, one was seen near Nantucket, and Captain David Thurlow of the schooner Lydia claimed to have harpooned one of two seventy-eight-foot serpents he saw six leagues off Mount Desert Rock, off the coast of Maine. The monster tore free and escaped.

In 1830, the action moved farther south. Captain Deland of the schooner *Eagle* had a close encounter with a sea dragon in Simon's Bay, South Carolina. The monster had a scaly, grey, humped, snake-like body and a head like that of an outsized crocodile. It had surfaced some nine hundred feet from his ship. He manoeuvred his vessel to within seventy-five feet of the monster, then shot at it with a musket. The monster dove and lashed the ships underside three times with its tail, causing the *Eagle* to rock violently. It was then joined by another, smaller serpent, and the pair swam away. Captain Deland thought that he had got off lightly, as he was convinced the monster could have wrecked his ship. This begs the question of why he shot at it in the first place. It might also answer the question of what has happened to some of the vessels that vanish without a trace. Perhaps they angered a less laid-back sea monster!

And so, the sightings continued. Even the British consul in Massachusetts saw it on two occasions from his hotel window! But we have examined quite enough here. Suffice it to say that from Nova Scotia right down the east coast of the United States, the many-humped sea serpent was seen almost every year until a sharp drop-off in the 1840s. Why was this?

It is interesting to speculate as to whether the global crash in whale populations had an adverse effect on sea serpents. If the serpents preyed on the whales, the monsters themselves would have experienced a population crash. Increased shipping may have also had an adverse effect on the many-humped sea serpents off the Atlantic coast of the United States.

They are still occasionally reported from these waters, but with nothing like the regularity of the sightings in the first half of the nineteenth century.

On the Pacific coast, the many-humped sea serpent has been reported since time immemorial and this continues unabated to this day.

The Native Americans have long known of sea dragons in the area and recorded them in rock art and carvings. The Chinook of British Columbia called it *hiachuckaluck*. The Hurons of Saint Lawrence Velley knew it as *angoub*. To the Manhousat of Flores Island and the Sydney Inlet it was *hiyitl'iik*.

The credible sighting by a non-Indian took place off the Queen Charlotte Islands in 1897. The description is by a prospector called Osmond Furgusson and was found in the British Columbia Provincial Archives by archivist David Mattison. It occurred on the morning of the 26th of June.

> *About 4.30 this morning we left Caedoo. I was steering the boat and pushing the oar at the same time. There was no wind. The boat was one hundred yards from shore, going south with a fair tide. I saw ahead of us what I thought was a piece of drift wood. On getting closer, I noticed it was moving towards us. When within fifty yards, I said to Walker (my partner), what is that? It seems to be moving this way. What we could see was an object sticking out of the water about two feet. When within a few feet of it the end uncoiled and raised a long neck about five feet out of the water with a head like a snake. The arched portion making a broad flat chest like I have seen on the cobra I think.*

> *When the serpent or whatever it was saw us, it turned slightly towards land to avoid the boat. The head and neck were almost immediately put underwater again. As it passed the boat, at a distance, that with an effort, I could have thrown an oar on it we could see a body about twenty-five feet long tapering with a fish-like tail and I think a continuous fin running the length of the body.*

> *A slow undulating motion went along the body, while the tail part had a steady sweep from side to side of about six feet. A curious thing was the broad neck or chest part that formed the arch.*

The only part out of the water when the head was down was not exposed broad-ways in the direction the fish was going, but had a decided twist to the left allowing the water to flow through it.

Though the animal is unlikely to be a fish, Fergusson's report amply illustrates the many-humped sea serpent's ability to flex both horizontally and vertically.

In 1917, R. M. Elliot saw a serpent whilst working on a telegraph line between Jordan River and Port Renfrew. It showed eight feet of head and neck plus seven humps. He shot at it from six hundred feet away. (What is this obsession people have with trying to kill anything that moves?) The animal reacted by thrashing violently but calmed down and swam on towards Victoria.

The animal was fired upon again in 1925 by Jack Nord. He was returning from Cape Mudge to Menzies Bay with his friend Peter Anderson. When abreast of Race Point, they noticed a huge animal resting on the surface of the water. Nord estimated it was one hundred to one hundred and three feet long and about two and a half feet in diameter. It had a saw-toothed fin on its camel-like head, and its mouth bore eight-inch fangs. Nord fired twice at it and missed. It did not seem bothered by the attack.

In 1932, Hubert Evens, Dick Reeves, and Bob Stevens saw the monster off Robert's Creek. They watched a series of humps break the surface, then an eight-foot neck with a horse-like head. The trio had a camera, but amazingly, no film!

FW Kemp, an officer of the Provisional Archives, gave the most detailed account. He made his report to the *Victoria Daily Times* in 1933.

On Aug 10, 1932, I was with my wife on Chatham Island in the Strait of Jaun de Fuca. My wife called my attention to a mysterious something coming through the channel between Strong Tide Island and Chatham Island. Imagine my astonishment on observing a huge creature with its head out of the water travelling about four miles per hour against the tide. Even at that speed a considerable wash was thrown against the rocks, that gave the impression that it was more reptile (i.e. lizard or saurian) than serpent to make so much displacement.

The channel at this point is about five hundred yards wide. Swimming to the steep rocks of the island opposite, the creature shot its head out of the water on to the rock, and moving its head from side to side, appeared to be taking its bearings. Then fold after fold if its body came to the surface. Towards the tail it appeared serrated with something moving flail-like at the extreme end. The movements were like those of a crocodile. Around the head appeared a sort of mane, which drifted round the body like kelp.

The thing's presence seemed to change the whole landscape, which makes it difficult to describe my experiences. It did not seem to belong to the present scheme of things, but rather to the Long Ago when the world was young. The position it held on the rocks was momentary. My wife and sixteen-year-old son ran to a point of land to get a better view. I think the sound they made disturbed the animal. The sea being very calm, it seemed to slip back into deep water; there was a great commotion under the surface, and it disappeared like a flash.

In my opinion its speed must be terrific, and its senses of smell, sight, and hearing developed to a very high degree. It would be terribly hard to photograph, as its movements are different from anything I have ever seen or heard of. I say its length to be not less than eighty ft. There were some logs on Strong Tide Island which gave me a good idea of the size of the monster as it passed them. I took a measurement of one the next day which was over sixty feet, and the creature overlapped it to a large extent at each end. I put a newspaper on the spot it rested and took an observation from our previous point of vantage. The animal's head was very much larger than the double sheet of newspaper. The body must have been at least five feet thick, and was of a bluish-green colour which shone in the sun like aluminium. I could not determine the shape of the head, but it was much thicker than the body.

Mr. Kemp had kept the sighting to himself but came forwards after another sighting that was made by Major H. W. Langley in the same area.

The Major and his wife saw a one-hundred-foot serpent as wide as a car, and with a serrated back, from their yacht.

The Major's sighting was the first to stir up much media interest. Inevitably the creature was given a name 'Cadborosaurus' (after Cadboro Bay) by Archie Wills, editor of the *Victoria Daily Times*. It was soon shortened to 'Caddy' and stuck.

One morning in 1933, two young friends went out duck hunting off Gowlland Head. Cyril Andrews and Norman Georgeson got an amazing view of Caddy feeding.

> *Thinking I might alarm Norman, I did not draw his attention to what I saw, so he came along and picked me up at the point from which we had shot the bird. From there we paddled to the wounded bird in the kelp bed. I was sitting in front of the punt ready to pick the bird up, when about ten feet away from it, out of the sea rose two coils. They reached a height of at least six feet above me, gradually sinking under the water again, when a head appeared. The head was that of a horse, without ears or nostrils, but its eyes were in front of its head, which was flat just like a horse.*

> *I attracted Georgeson's attention to it and he saw one coil and the head well clear of the water. Then the whole thing, except the head, which remained just out of the water, sank. I was still only ten feet away from it, with the duck right beside the thing, when to my horror it gulped the bird down its throat. It then looked at me, its mouth wide open, and I could plainly see its teeth and tongue, which were those of a fish. I would swear to the head being three feet long and two feet wide. When it closed its mouth, all the loose skin folded in neatly at the corners while its breathing came in short, sharp pants, like a dog after a run. At that point a number of sea gulls swooped down at the creature, which snapped at them when they came too close. Shortly after this it sank beneath the surface.*

The friends called Justice of the Peace G.F. Parkyn, who took down an affidavit of what they had seen. Then the monster surfaced again eighteen metres from shore. Eleven other people including Mr Parkyn saw it. Cyril, and two friends—Kathleen Georgeson and Arthur Pender—saw the animal twice again. On one occasion they saw it snap at sea gulls once more.

This duck-eating and water bird-stalking has been observed in the behaviour of Ogopogo at Lake Okanagan. Could the animals be related? Sea serpents have been seen swimming up rivers on many occasions. They may well be the same as some types of lake monsters.

The many-humped is no stranger to the UK shores. Many have been seen around the British Isles, often off the east coast of England. Mr George Ashton, a forty-nine-year-old shot blaster from Sheffield, and his wife, May, saw such a beast three hundred feet from shore at Chapel St Leonards in October 1966.

It had a head like a serpent, and six or seven pointed humps trailing behind. When I have been out at sea, I have seen seal and sea snakes swimming about and what I saw was neither of these. At first, I thought it was a log but it was travelling at about eight knots and going parallel with the shore. We watched it for some time, coming from the direction of Chapel Point, until it disappeared out of sight towards Ingoldmells. I just didn't believe in these things and tried to convince myself it was a flight of birds just above the water. I even thought of a miniature submarine, but after watching it for some time, I knew it couldn't be.

If the word of a shot blaster isn't good enough for you, then how about the daughter of one of our country's greatest novelists? On July 20, 1912 Miss Lilias Haggard, daughter of Sir Henry Rider Haggard, author of *King Soloman's Mines*, was privy to a sighting that could have slithered from the pages of one of her father's novels. She wrote to him from her home, Kessingland Grange, East Anglia, to tell him of the encounter:

We had great excitement here this evening. And we are convinced we saw a sea serpent! I happened to look up when I was sitting on the lawn, and saw what looked like a thin, dark line with a blob at one end, shooting through the water at such a terrific speed it hardly seemed likely that anything alive could go at such a pace. It was some way out over the sandbank, and travelling parallel with the shore. I tore into the morning room and got the glasses, and though it had, at that moment, nearly vanished in the distance, we could make out it had a sort of head at one end and then a series of about thirty pointed blobs, which dwindled in size as they neared the tail. As it went along it seemed to get more and more submerged, and then vanished. You can't imagine the pace it was going. I suppose it was about sixty feet long.

Her father sent the letter to the *Eastern Daily Press* along with a letter from himself asking if anyone else had seen the creature. A number of people responded. Mr. C. G. Harding said that he has seen saw a long, dark creature moving through the water like a torpedo the day after Miss Haggard's sighting. Mrs Adelaide J Orams and her son had seen a dark object swimming out to sea at Mundesley. An anonymous woman claimed to have seen it three weeks previously, moving with 'lightning rapidity' opposite the harbour mouth at Gorleston. Mr W. H. Sparow and his wife had seen it the day before the Haggard sighting from the promenade at Cromer. It was moving at forty miles per hour, and undulating. He estimated it to be thirty feet long.

The Long-Necked Sea Serpent

Now we must turn our attention to the other main type of sea serpent, the long-necked. The long-necked sea serpent—as its name suggests—has a snaky, elongated neck. Its body is shorter than the many-humped but considerably wider. The animal has two sets of flippers attached to the barrel-shaped body, as well as a tail at the rear. Overall it superficially resembles the extinct marine reptiles known as plesiosaurs. Some confusion may arise between the many-humped and the long-necked. The body of the long-necked can show up to five humps, but never as many as the many-humped. The humps of the long-necked seem to be flexations in the back or possibly fat storage as in camels. The humps of the many-humped would appear to be the loops of its elongated body showing as it propels itself. When rearing up with the front portion of the body, the many-humped can seem to have a long neck. The long-necked however really does have a long neck, quite separate from the bulbous body. Sometimes a dorsal fin is reported on the back of the long-necked, but these reports are in the minority.

The long-necked has been reported in all the world's seas. What follows is a smattering of reports to give the reader a feel for the animal.

Scientists have in the main poured scorn on sea serpents, but in 1905, two well-respected zoologists had a sighting of their own. E. B. G. Meade-Waldo and M. J. Nicoll were fellows of the Zoological Society, and well known in ornithological circles. The friends were on a zoological research cruise on board Lord Crawford's yacht *Valhalla*. They were fifteen miles out from the mouth of the Parahiba in Brazil. Meade-Waldo writes...

> *On Dec 7, 1905, at 10.15 a.m. I was on the poop of the 'Valhalla' with Mr Nicoll, when he drew my attention to an object in the sea one hundred yards from the yacht; he said; 'Is that the fin of a great fish?'*
>
> *I looked and immediately saw a large fin or frill sticking out of the water, dark seaweed-brown in colour, somewhat crinkled at the edge. It was apparently about six feet in length and projected from eighteen inches to two feet out of the water.*
>
> *I got my field glasses on it (a powerful pair of Goerz Trieder), and almost as soon as I had them on the frill, a great head and neck rose out of the water in front of the frill; the neck did not touch the frill in the water, but came out of the water in front of it, at a distance of certainly not less than eighteen inches, probably more. The neck appeared to be the thickness of a slight man's body, from seven to eight feet of it was out of the water; head and neck were all about the same thickness.*
>
> *The head had a very turtle-like appearance, as had also the eye. I could see the line of the mouth, but as we were sailing pretty fast and quickly drew away from the object, which was going very slowly. It moved its head and neck from side to side in a peculiar*

manner: the colour of the head and neck was dark brown above and whitish below—almost white, I think...

Another finned long-necked sea serpent was seen in Australian waters, showing just how far these animals range. The steamer Saint-Francois-Xanier was on the Tongking-New Caledonia-Australia run when she encountered a sea dragon. Her captain, Raoul Jaillard, recorded the sighting.

Haipong, 18 March 1925,

Sir, I am sending you a little sketch drawn at sea several minutes after the appearance of the famous sea serpent. The second captain, the second lieutenant, the radio officer, and the third engineer are unanimous in confirming the following lines:

On 2nd February 1925, while on passage from Noumea to Newcastle, the ship was making ten knots, at 18.30 hours abeam of Port Stephens on the east coast of Australia, two masses like turtle's shells were seen floating thirty feet from the ship on the starboard bow.

Abeam of the engines there rose a big head like a camel's head, on a long flexible neck having a great similarity to a swan's neck. The height of the neck was about eight feet. The body, thick as the big Bordeaux barrels, formed a chain of five loops; on the fourth loop, an aileron as on sharks of large dimensions, measuring five feet in height and in width at the base. The aileron seemed to be black in colour; the colour of the animal was dirty yellow, the skin smooth without appearance of scales.

As it passed astern of the ship and was abeam of the starboard screw, the animal's head began to move backwards and forwards, which led us to think it had been touched by a blade of the screw; its movement seemed hindered and it was not at all like that of the very little snakes seen near the coast.

The animal was visible for fifteen minutes, no optical illusion is possible. For besides the testimony of the Europeans, the Blacks from New Caledonia serving as seamen on board, the Annamite boys and the Chinese stokers all gave one cry: 'There is the Dragon!' The Chinese even made an offering to it.

As night falls very quickly at that time of year we could not give other details, being one and all fairly taken aback by this fantastic apparition...

RAOUL JAILLARD

The next encounter I am recounting is one unparalleled in sea serpent history. Despite their large size and massive strength, they do not seem to look on humans as a food source. But if one Edward Brian McCleary is to be believed then the long-necked sea serpent sometimes counts the 'long pig' as his prey.

The story begins on a sunny Saturday morning; March 24, 1962. Cleary, together with four friends, Eric Rule, Warren Sulley, Brad Rice, and Larry Bill, intended to go skin diving. They planned to dive on the wreck of the *Massachusetts*, several miles off the coast of their Florida home in Pensacola. The youths were all aged between fourteen and sixteen. They had a seven-foot Air Force life raft to transport them to and from the wreck. A seemingly flimsy craft for five teenagers, but the weather forecast had been good.

As they rowed out to the ship, they found that the tides were stronger than they had anticipated. Storm clouds were beginning to gather on the horizon. Three of the boys started to swim behind the raft, passing it onwards. They saw a boat and signalled for help, but it did not seem to see them. The group saw a buoy about a mile distant, and decided to head for that. As they paddled towards it the waves grew higher and started to swamp the raft.

They managed to get to the large buoy and clambered onto its metal scaffolding. Their raft was dragged under, as the buoy began to sway violently in the wind. The skies became black, and icy rain lashed down on the stricken divers.

After a while, the rain slowed and became a fine mist. The mountainous seas calmed to a mill-pool stillness, then a thick fog rolled in. Through the fog the boys heard a splashing sound. A foul odour like dead fish permeated the area. Then, silhouetted in the fog, the boys saw something like a telegraph pole rise up from the water, and dive back in. A strange whining cry filled the air. That was the last straw. The boys panicked and dove into the water. In the cold dark water, they heard the sea dragon hissing and splashing.

Warren called out:

'Hey! Help me! It's got Brad! It got Brad! I've got to get outta here.'

His scream was abruptly cut short in the dark. The three remaining boys tried to cluster together. Brian, Larry, and Eric found each other. They tried to swim away through the fog, but the seas were once more getting rough. Suddenly Larry vanished without a trace. Brian wrapped the exhausted Eric's arm around his neck for support, but a wave pulled them apart. Then, next to the stricken Eric, the monster's head and neck emerged. It had a face like a turtle's but more elongated, with green eyes. The ten-foot neck was a greeny brown.

It opened its jaws, bent over Eric, and dragged him under.

Brian, unsurprisingly, recalled little after this. He swam through the dark, cold sea expecting death at any moment. He sank and felt the tranquillity of death over take him.

He awoke on a beach some miles from Pensacola on Sunday morning with no memory of how he got there. He was found by a group of children. He was taken to Pensacola Naval Base Hospital suffering from shock, exhaustion, and exposure. When the Director of the Search and Rescue units, E. E. McGovern, came to see him McCleary told the whole story.

The bodies of Eric, Warren, and Larry were never found. Brad's washed ashore a week later. I have no idea what the outcome of his post mortem was. Brian understandably suffered a breakdown, and in later life became an alcoholic and now refuses to talk of the events.

What did happen to those boys on that night? Did Brian hallucinate the whole event through fear or guilt at his survival? Or did his friends fall victim to an unknown carnivorous sea beast? It seems unlikely that he would have made up such a story about the death of his four best friends. We shall never know. The last word on the case is best left to E. E. McGovern, who when visiting Brian in hospital said:

> *The sea has a lot of secrets. There are a lot of things we don't know about. People don't believe these things because they are afraid to. Yes, I believe you. But there's not much else I can do.*

The long-necked sea serpent is no stranger to the British coast. The largest concentration of sightings has been around Cornwall where the monster is known as 'Morgawr', (which allegedly means 'sea giant' in archaic Cornish). Harold T. Wilkins and a friend saw two specimens whilst at the shark fishing port of Looe in 1949.

> *Two remarkable saurians nineteen to twenty feet long, with bottle green heads, one behind the other, their middle parts under the water of the tidal creek of Looe, east Cornwall, apparently chasing a shoal of fish up the creek. What was amazing were their dorsal parts: rigid, serrated, and like the old Chinese pictures of dragons. Gulls swooped down towards the rear of one. These monsters, two of which were seen, resembled the plesiosaurus of Mesozoic times.*

Forty-nine years later, the *Falmouth Packet* newspaper reported the first in a whole series of modern sightings. Two witnesses, Mrs Scott and Mr Riley, had seen the thing off Pendennis Point. They described a long neck and a small head furnished with stubby horns. The neck had what looked like a mane of bristles running along it. The monster dived and surfaced holding a large conger eel in its mouth. Mrs Scott commented that she would never forget the face on the thing as long as she lived.

On the 28th of December 1975, Mr Gerald Bennett wrote to the same paper about his own sighting.

> *I myself, during the last Christmas holidays, witnessed the sighting of a similar creature (to that seen by Mrs Scott and Mr Riley), although until now I have remained reticent about it. It was off the shore at Durgan, Helford, about 4 p.m., near dusk. When I first spotted it, I thought it was a dead whale, but as I drew nearer it started to move away, smoothly, and I could see it was not a whale, nor like any creature seen round here. I judged that the*

part of it I could see above water was about twelve feet in length with an elongated neck.

In January 1976, Duncan Viner, a dental technician, saw a forty-foot monster off Rosemullion Head. He too thought it was a whale until a long neck emerged from the water.

Later the same month, Amelia Johnson saw it in the same area, and described it as:

A sort of prehistoric, dinosaur thing, with a long neck, which was the length of a lamppost.

Sightings continued. Two London bankers, Tony Rodgers and John Chambers, were fishing on the rocks of Parson's Beach at the mouth of the Helford River, when they saw Morgawr. It was green-grey in colour and bore humps. Rodgers thought he saw a second smaller creature accompanying the first.

In the summer of 1976, George Vinnicombe and John Cock were fishing the war time wrecks twenty-five miles out from Lizard Point! Once again, a dead whale was thought to be the object they saw floating on the calm surface. The idea was quashed as a serpentine head and neck rose up before the monster dived.

Brother and sister Allan and Sally Whyte came upon the monster on land. The brown coloured twenty-foot animal was resting on Grebe beach and slithered into the sea at their approach.

In July of 1985, two girls, Jenny Halstead and Alice Lee, from the lovely Yorkshire town of Hebden Bridge, spotted the beast whilst on a cycling holiday.

At some time between 6.30 and 7 p.m., from a position at Rosemullion overlooking the sea, we witnessed a genuine living monster of the deep, which we believe must be your legendary monster Morgawr. The creature's back broke the surface and looked rather like a massive overgrown black slug. We both watched the animal for about ten seconds as it wallowed in the water. Then the creature sank beneath the waves and did not surface again. Even though we had a camera to hand, we were too astounded by the sight by the sight of the monster to think of taking a photograph until it was too late.

Josh Tomkins, a fisherman, and his son were out in a boat a mile off Falmouth on 24th of August 1999 at 4.30 p.m. when they saw something rise from the water.

Initially I thought it was a dead body rising to the surface. As we watched the mound, it dropped back under the water, causing a terrific swell. Moments later it resurfaced, about fifty yards from our boat. I could see that it was no dead body, but a large creature. My son thought it was the back of a whale as this was the most logical explanation we could find. Our opinions altered when, about ten yards in front of the mound, a small head appeared above the surface. The head lifted out of the water only very slightly but sufficient for both of us to see part of what seemed a long slender neck. It then dropped back down in a colossal disturbance.

We were both shocked by the immense size of the creature; it was like no fish I have ever seen, in fact it wasn't like anything I had seen before. I am pretty sure it must have been Morgawr the sea monster. I didn't believe in this before and I am still not certain now, but that thing sure did look like a dinosaur-like creature. I would think it was dark brown or black, but the colouration was not evenly distributed, it seemed to be patchy in parts, slightly lighter in areas. We both saw its eyes, no ears and no mouth. It made no noise, just created a huge wash as it submerged. After seeing it, I would not be too happy about going out into open water after dark in a small boat; it's very large and could inflict some serious damage to a small vessel.

Mrs Elsie Morgan saw it at around about the same time in the waters off Falmouth.

I saw, about one hundred yards seaward, a black object appear in the water. It appeared before my eyes and seemed to be stationary. I would estimate that it was about ten feet long and, at its highest point out of the water, about two feet. As I watched, something rose out of the water close to the hump-like mound. It appeared to rise to an angle of forty-five degrees and looked a bit like the curved end of a question mark, but more angular. I then realised that this was either the tail or the head and neck of some large marine animal. This dipped in and out of the water several times, its highest point appeared to be moving from side to side, like the head of a snake looking around. It remained in sight for a minute or two before sinking from view. I could see white foam on the sea surface where it disappeared. It wasn't like any sort of animal or fish I had ever seen, nor could I imagine what it looked like as a whole, but it was very large and looked quite cumbersome.

The most recent sighting to date occurred off Falmouth again on May 16, 2000. Derek and Irene Brown had parked their caravan overlooking the sea.

The sea was quite calm, not choppy or heavily disturbed, and the weather was reasonably good, by that I mean that no mist or rain was falling, and visibility was clear for a considerable distance out to sea. As we sat next to our caravan overlooking the sea, I saw something appear in the water perhaps two hundred yards away, certainly no more. I took no notice of the object as the sea does throw up debris and bits of driftwood and I had no reason to concentrate on the object. I looked away and heard Irene ask me 'What is that out there?' I looked again as she pointed to the object I had glimpsed a few moments earlier. The object now took the form of defined humps, two of them very close together. I would think that overall, they measured about fifteen feet. I estimated that from my height, I am just over six feet tall. The humps were still, and as I sat searching for an explanation to give Irene, a periscope-like object came out of the water very close to these humps. It was moving in a flexible manner, not at all rigid. I would think it looked close to the stance a cobra or python makes, raising its head and neck before it strikes. Irene shouted, 'It's an octopus,' but it clearly wasn't. I took the humps to be the back of a large body, the periscope-like object being the head and neck. I told Irene that I thought it was a monster and to get the camera from the car as we should take a picture of it.

As she got up to leave me, the creature seemed to roll forwards, dipping head first into the water. There was a huge commotion as it disappeared. Irene came back with the camera but it had gone. We stayed to look for the creature for another hour, but it never resurfaced.

I cannot begin to explain how we felt about what we saw. We decided to keep it to ourselves, as no one would believe us, and we would look stupid. I think the creature you are looking for is not one known to zoological science, but more to archaeologists who search for fossilised remains of creatures that existed many millions of years ago. This may sound stupid and far-fetched, but somehow, I believe that some of them lived on and exist in our waters. It wasn't a fish, more like a water-based dinosaur, like something you see in those 1960 films about prehistoric times.

I am not a storyteller, not do I wish to capitalize upon what my wife and I saw, but I felt I should report this to someone, as it genuinely happened.

Almost every county in Britain with a coastline has had reports of long-necked sea serpents. In August 1963, Mr P. Sharman was on holiday in Wales when he saw one from his vantage point on the cliffs. He was near New Quay, Cardigan Bay. He wrote to the late Tim Dinsdale, a legendary investigator of lake and sea monsters, with the details.

I noticed an animal greatly disturbing a colony of seals. The creature drawn was slowly moving its four paddles to and fro as if in readiness to make a sudden move. At one end of it there appeared to be a long neck and small head poised above the water as if to strike out suddenly. The seals around it were making off as though the fear of death was upon them. This led me to suspect that the creature was making ready to kill a seal. After I had watched the thing for a few minutes I realised there was a remote possibility that I was looking down upon a floundering basking shark. This seemed more and more probable, so I left the scene.

Later, during that week I was exploring another cove about half a miles from the spot where I saw the strange animal. Here I saw the carcass of a seal with a huge chunk bitten off from its neck and shoulders. This practically cut the body in two and I could not help wondering what creature could have made such a horrible wound. Of course, it could be that I saw a basking shark half in and half out of the water and mistook the tail for the head and neck of a Plesiosaur-type creature. But I saw no dorsal fin; and are basking sharks aggressive to seals? The creature, comparing it with the seals must have been thirty to forty feet long, and was a brownish black in colour. I was looking down at it from about one hundred feet at an angle of fifty degrees. It must have been eight feet wide.

Mr Sharman provided a drawing with his statement. It shows a large long-necked animal with a bulbous body, two pairs of flippers, and a stout tail. All around, seals are scattering from it.

Filey Brig is a long, low spur of rocks jutting a mile out to sea from the coast of the Yorkshire seaside town of Filey. In local legend they are said to be the bones of a dragon. Fittingly it was here that one of the spookiest encounters with a long-necked sea serpent occurred. In February 1936, Wilkinson Herbert, a coastguard, was walking along the Brig on a dark moonless night...

Suddenly I heard a growling like a dozen dogs ahead, walking nearer I switched on my torch, and was confronted by a huge neck, six yards ahead of me, rearing up eight feet high!

The head was a startling sight, huge; tortoise eyes, like saucers, glaring at me, the creature's mouth was a foot wide and its neck would be a yard round.

The monster appeared as startled as I was. Shining my torch along the ground, I saw a body about thirty feet long. I thought, 'this is no place for me' and from a distance I threw stones at the creature. It moved away growling fiercely, and I saw the huge black body had two humps on it and four short legs with huge flappers on them. I could not see any tail. It moved quickly, rolling from side to side, and went into the sea. From the cliff top I looked down and saw two eyes like torch lights shining out to sea three hundred yards away. It was a most gruesome and thrilling experience. I have seen big animals abroad but nothing like this.

So what manner of beasts are the sea serpents? The most popular theory is that they are decendents of prehistoric marine reptiles. In recent years this idea has gone out of fashion in favour of unknown marine mammals. I think that this is very premature. Some sea monsters are undoubtedly marine mammals, but it is my belief that the kinds tackled here seem to be true reptiles.

The plesiosaurs have been touted as favourites in the past. Indeed, the long-necked bears a striking resemblance to the plesiosaurs, however they belong to a family of reptiles extinct for sixty-five million years. They died out at the end of the Cretaceous period along with non-avian dinosaurs. But there were other marine reptiles that belong to or are very closely related to families living today.

The mososaurs were elongated crocodile-like predators with four flippers and savage jaws. Some, such as tylosaurus, grew to forty feet or more in length. They were closely related to the varanids or monitor lizards that thrive today.

Another possible sea dragon ancestor was a group of prehistoric marine crocodiles known as thalattosuchians. Some of these bore fins rather than feet, and some such as *steneosaurus* and *metriorychus* were capable of vertical flexation, a feature reported in the majority of sea serpent cases.

Sixty-five million years of evolution could have adapted such creatures to cope with cold as well as warm water. It could also change their body shape radically; perhaps to compete with the emerging marine mammals.

CHAPTER FIVE

Giant Apes and Hominins

'I was walking in the mountains when I
saw a wild beast. As it drew closer, I saw
it was a man. As it drew closer still, I saw
it was my brother.'

—Indian proverb

•

All over the world there are reports that stretch back centuries, of encounters with man-like, hair-covered beasts that walk on two legs. The *Epic of Gilgamesh*, written in 2100 BC, includes a character called Enkidu who is a hairy, powerful wildman who dwells in the wilderness, knowing nothing of civilization.

In this chapter we shall examine such creatures and the new fossil and DNA evidence that supports their existence.

A quick note on terms. Hominin refers to modern man and man's pre-historic relations. Hominoid refers to all of these *and* the great apes.

Asia

The most famous of the man-beasts is undoubtedly the Yeti. It is also the most misunderstood as well. In the mind of the general public, the Yeti is a white-furred beast, a kind of hybrid of a polar bear and a gorilla that prowls the eternal snows of the Himalayas. In fact, the Yeti is not white. It's hair ranges from reddish to brown to black. In all my years of research, I have only come across two reports that give the Yeti white hair. The confusion comes from a mistranslation of one of the beast's many names, *metoh-kangmi*, Sino-Tibetan for 'abominable man of the rocks'. It was mistranslated as 'abominable man of the snows'. This is where we get the western term abominable snowman. It is also where the false idea of a white, snow-dwelling beast comes from. Above the snowline, there is little for a large primate to eat. The lush forests in the lower valleys make much more sense for a Yeti inhabit. The term Yeti is Tibetan for 'rock beast'.

Another fact the public doesn't realise is that the term Yeti is applied to three different creatures. The *Dzu-teh* is a hulking biped eight to ten feet tall with dark hair. It leaves massive, manlike foot prints. The *Mi-teh* is more man-sized and moves both bipedally and on all fours. It has reddish hair and leaves tracks that have a divergent big toe. The smallest type, around four feet tall, is known as the *Teh-lma* and has light brown to yellowish hair. The creatures have many regional names and are reported from the Himalayas, Tibet, China, Malaysia, and India.

The pre-Buddhist Himalayan people spoke of a huge, vastly strong, hairy, manlike beast of the forests and mountains. It used rocks as weapons and made a whistling sound. This ancient description fits in well with the reports of modern times.

British naturalist and explorer Brian Houghton Hodgson spoke of his guides fleeing from a huge, hair-covered, upright, ape-like beast. Hodgson thought it might have been an orangutan. In modern times, though, these have been confined to Borneo and Sumatra.

Indian Army surgeon, Laurence Waddell, was told by his guides in Tibet of huge ape-like creatures. The guides described the beasts' tracks but Waddell dismissed them as bears.

Western interest did not really peak until 1951 when English mountaineer Eric Shipton took an iconic photograph of a strange footprint he discovered on the Melung Glacia in Nepal. The famous picture shows an ice axe next to the print for scale. The twelve-inch-long prints sunk deep into the snow and showed an offset big toe like that of an ape.

Two years later, Edmund Hillary and Tenzing Norgay saw massive ape-like tracks in the same area. Norgay said he thought that the Yeti was a type of ape and that his father had seen the creature twice.

The most famous account of a run in with the larger type of Yeti occurred during WWII. Sławomir Rawicz, a Polish soldier who, after the Nazis and the Soviet Union had defeated Poland, was arrested and sent to a Siberian prison camp, charged with being a spy. He and six companions escaped the gulag in 1941 and began a trek, on foot, to India.

He later recounted his adventures in the book *The Long Walk*. In 1942, the group were passing through a mountain valley in Tibet when they saw two figures ahead of them.

Two points struck me immediately. They were enormous and they walked on their hind legs. The picture is clear in my mind, fixed there indelibly by a solid two hours of observation. We just could not believe what we saw at first, so we stayed to watch. Somebody talked about dropping down to their level to get a close-up view. Zaro said, 'They look strong enough to eat us.' We stayed where we were. We weren't too sure of unknown creatures which refused to run away at the approach of men. I set myself to estimating their height on the basis of my military training for artillery observation. They could not have been much less than eight feet tall. One was a few inches taller than the other, in relation of the average man to the average woman. They were shuffling quietly round on a flattish shelf which formed part of the obvious route for us to continue our descent. We thought that if we waited long enough, they would go away and leave the way clear for us. It was obvious they had seen us, and it was equally apparent they had no fear of us.

The American said that eventually he was sure we should see them drop on all fours like bears. But they never did. Their faces I could not see in detail, but their heads were squarish, and the ears rested close to the skull because there was no projection from the silhouette against the snow. The shoulders sloped sharply down to a powerful chest. The arms were long, and the wrists reached the level of the knees. Seen in profile, the back of the head was a straight line from the crown into the shoulders—'like a damned Prussian,' as Paluchowicz put it. We decided unanimously that we were examining a type of creature of which we had no previous experience in the wild, in zoos or in literature. It would have been easy to have seen them waddle off at a distance and dismissed them as either bear or big ape of the orangutan species. At close range they defied facile description. There was something both of the bear and the ape about their general shape, but they could not be mistaken for either. The colour was a rusty kind of brown. They appeared to be covered by two distinct kinds of hair: the reddish hair which gave them their characteristic colour forming a tight, close fur against the body; mingled with these were long, loose, straight hairs, hanging downwards, which had a slight greyish tinge as the light caught them...

We pushed off around the rock and directly away from them. I looked back and the pair were standing still, arms swinging slightly, as though listening intently. What were they? For years they remained a mystery to me, but since recently I have read of scientific expeditions to discover the Abominable Snowman of the Himalayas and studied descriptions of the creature given by native hillmen. I believe that the five of us that day may have met two of the animals. If so, I think recent estimates of their height as about five feet is wrong. The creatures we saw must have been at least seven feet.

Some of Rawicz's companions were too scared to carry on and the group took another, more dangerous route that caused some of them to fall to their death.

Boris Porshnev, the Russian hominologist, recorded a sighting of a huge Yeti from the village of Tarka in Nepal in 1958.

As one of the farmers was going to the mill in the morning, he saw that the door was open and he peered inside...

But I did not see a man. It was a huge humanoid creature, all covered with thick long hair. It was eating all the flour and grain it could find. The creature was about three metres tall, and the arms were long enough to touch the knees. It was intent on searching for food, it did not even notice me. I have heard many stories about these creatures, and know it brings bad luck to meet one, but I still decided to look at it properly. It had a flat bald face with lots of wrinkles like a monkey. The head was high and conical, and the body was covered with hair. It had no tail. The nails were like claws, and it kept roaring. The whole body of the creature was white with flour. The Yeti had not noticed me, so I walked carefully away, and ran to the village to get others to help. But the Yeti heard me, and started climbing towards the snow-covered mountain peak. The villagers who heard my screams, saw him running away.

Three villagers confirmed the story.

The Texas oil millionaire Tom Slick financed a number of expeditions to search for the Yeti in the late '50s and early '60s. He was told the following concerning an incident on the road to the Barun Valley, Nepal, by the witnesses themselves. A Yeti had been seen just two days before, and the two eyewitnesses, a boy and his older sister, were interviewed separately. They were grazing their herd of yaks at an altitude of fourteen thousand seven hundred feet. Sometime in the afternoon, at large upright animal came out of the bushes and approached them to a distance of about one hundred feet. The boy said the animal was about ten feet in height. According to both, the creature was covered with black and brown fur, except for the waist, where the fur was whitish. The head was high and conical. Both eyewitnesses were shown various pictures of other animals as well as reconstructions of the Yeti and of prehistoric man. Both said the last two were most accurate. They also noted some similarities to a photo of an orangutan, but rejected pictures of black bears, monkeys and other Indian animals. Perhaps the most interesting aspect of the case is the fact that, after having run away in fear, the two shepherds returned to their animals. The Yeti had disappeared, but one of the yaks was lying on the ground with a broken neck, as if from a blow.

In 1968, Naiqiong, a hunter from Chekaung Village, Chenkaung District, Dingjie County, East of the Himalayas, was staying in a mountain cave overnight. The cave was at a site known as Lu'ema. Naiqiong had built a large fire in the entrance to keep animals at bay. Awakened by noises in the night he found two hair-covered, manlike creatures hurling rocks at him from beyond the bonfire. They were larger than humans and had greyish hair. He grabbed his gun ready to shoot them. The hunter and the strange beasts stood looking at each other until dawn when the Yetis moved away.

Though smaller than the Dzu-teh the Mi-teh is supposed to be quite aggressive. A Yeti was alleged to have attacked Lhakpa Domani in 1974 near Mount Everest as she tended her yaks. She heard a whistling noise and then was struck from behind. She described her attacker as similar to a large ape with black and brown hair and broad cheek bones. It was said to have picked her up and thrown her some distance, then attacked the yaks that she had been tending. Her brother found her soon after, wounded and unconscious, but alive. Several nearby yaks lay dead, half eaten. The Yeti's footprints were all around them. Later, an investigating police officer found the creature's tracks and saw the dead yaks.

Legendary explorer Peter Byrne interviewed a man in the Arun Valley of Nepal who was the foreman of a group of workers. The man told Byrne that in 1956 the workers had panicked when they saw a Yeti emerge from the forest close to where they were working. It was five feet tall and covered in black hair. Hearing their screams, the creature turned and walked back into the forest. The man told Byrne that the people of the Arun feared the Yeti because it would attack a lone man in the forest. For that reason, they travelled in groups. The incident happened five or six miles north of the town of Tumlintar.

Another man told Peter Byrne that many years before his father had been attacked by a Yeti whilst on a pilgrimage with friends to Manasowar Lake in Tibet. On a hillside south of the lake, a Yeti charged at the group, coming down the slope with great strides. It was covered in hair, upright like a man, leaping over the rocks and uttering shrill cries. The party all fled.

Explorer and conservationist Desmond Doig visited Nepal in 1960 and wrote to Boris Porshnev of his findings.

> This morning I met two people who saw the Yeti close enough and for enough time to give a detailed description. Twenty years ago, the village of Beding, two hundred kilometres from Kathmandu, was covered by a large avalanche. Some of the people were buried alive in their stone huts. It was in that year, one of the residents remembered, that the Yeti made themselves felt very strongly.

They made a terrible whistling in the cold winter nights, and every morning their tracks could be found in the fresh snow.

The head lama of the local small monastery told Doig, that he was certain the Yeti were looking for the bodies of the victims of the avalanche. Doig continued.

Just two winters ago, a pair of 'snow people' came up to the monastery at the time of the evening prayer. It was already dark and snowing. They circled the building and even tried to break through a window. The monks were very frightened (there were only five people in the monastery) and started ringing the sacred bell as much as they could. That made the 'Yeti' run away howling. Their voices were very similar to human voices, and the sounded full of sorrow. The abbot of the monastery told about another event, that took place several years before. A Nepali official with servants came to Beding in the winter to hunt deer and pheasants. The group had arrived and was about to go to bed, when they heard the noise of several large animals circling the building. Suddenly a large head overgrown with hair appeared at the window. Nobody dared to move to grab a gun. The Yeti howled with rage. The lama heard the noise of the animal, and started to blow the big horn. The frightened animal fled, making a terrible noise.

In 1972, biologists Edward Cronin and Jeffrey McNeely found a set of ape-like tracks outside of their camp in Arun Valley. Something had come down from a ridge line and investigated their camp at night. It seemed to walk on two legs. The tracks were fresh and the morning sun had not yet distorted them. After investigating the camp, the tracks went back up to the ridge without dropping onto all fours. They vanished into rhododendron thickets. The pair made casts of the prints that were later viewed by mammologist Dr George Schaller who pioneered studies of gorillas. He noted that they demonstrated a close resemblance to the mountain gorilla. Cronin felt that they lend credibility to the idea that the Yeti was an unknown form of hominoid.

So, what is the Yeti? Some have suggested that it is nothing more than a bear. Italian mountaineer Reinhold Messner claimed in his 2000 book, *My Search for the Yeti*, that the creature was nothing more than a brown bear. This is curious as on previous occasions he had claimed to have seen the Yeti and described it as a primate-type animal.

I once interviewed the actor Brian Blessed, a renowned explorer and mountaineer himself, for a long defunct and not very good magazine called *Quest*. Blessed, who is a friend of Messner, said that he had told him of his encounter with a Yeti. Blessed said that Messner had walked around some rocks and come 'face to face' with the creature. He said it was not a bear, but was seven feet tall, manlike and stood erect.

There are other occasions when Messner's descriptions sound precious little like a bear. Julian Champkin of the *Daily Mail* wrote on August 16, 1997 that Messner had...

> ...encountered the Yeti; and not once, but four times, once close enough to touch it. More importantly, he claims to have photographs of the creature, including a mother Yeti tending her child, and a Yeti skeleton.

Needless to say, none of his pictures have been forthcoming. Messner goes on:

> ...I searched for a week, twelve hours a day, in an area with no trees. I didn't expect to find one so soon. First, we saw a mother with her child. I could only take a photograph from the back. The child had bright red fur, the older animal's fur was black. She was over two metres tall, with dark hair, just like the legend. When they saw us, they disappeared.

Two days later, he claimed to have come across and filmed a sleeping Yeti. The film is just as noticeable as the photos by its absence.

In an article relating to the BBC's *Natural World* documentary on the Yeti, Messner describes seeing one from a range of thirty metres in Southern Tibet. The article says Messner is sure it is some kind of primate. He describes it in the article thus:

It was bigger than me, quite hairy and strong, dark brown-black hair falling over his eyes. He stood on two legs and immediately I thought he corresponded to the descriptions I heard from Sherpas and Tibetans.

So why did Messner write a book trying to explain away the Yeti as a bear when this transparently was not the creature he claimed to have seen? Was it because of fear of ridicule? And what became of the photos and film? Was Messner trying to take the focus away from these or make them seem less important by saying the Yeti was just a bear? Could this be because the film and photos did not exist?

Sherpas become angry when westerners say that the Yeti is just a bear, and quite rightly. The animal they pick repeatedly as looking most like the Yeti is the gorilla, but walking on two legs rather than four. The Yeti has a flat, ape-like face. The Yeti walks almost constantly on two legs. The Yeti can manipulate things with its hands and hence must have opposable thumbs. It is said to sometimes hurl large rocks. Bears have none of the above features. The Yeti is clearly some kind of primate, most likely a great ape. Until he delivers the goods, I'm inclined to dismiss Messner's claims.

The prime candidate for the larger kind is a massive ape from the fossil record known as *Gigantopithecus blacki*. The creature is known only from its massive fossil teeth and jaws. The fossil teeth were first found in a Chinese apothecary shop in 1935 by Dutch palaeontologist Gustav Heinrich Ralph von Koenigswald. They were being sold as 'dragon's teeth'. Koenigswald recognised them as the molars of a titanic ape. Post-cranial remains have never been found but extrapolating from the size of the teeth and jaws Gigantopithecus may have stood ten feet tall and weighed one thousand three hundred pounds. The flaring of the lower jaws made Grover Krantz and Jeff Meldrum conclude that the neck extended directly under the creature's head, meaning that it walked upright on two legs.

Pitting and wear patterns on the teeth of Gigantopithecus suggest a fibrous diet similar to that of the giant panda. The creature probably fed on bamboo but fossilised seeds found lodged between the teeth prove that it also fed on fruit. Gigantopithecus fossils have been found in China, Vietnam and India. It was a hugely successful primate species existing for over two million years before becoming extinct one hundred thousand years ago due to climate change. However, some think that the animal simply retreated into the mountain forests and still exists today.

The medium-sized Yeti seems to be a great ape. It moves on two legs but can clamber in trees and bound on all fours. In the Pleistocene era 2,588,000 to 11,700 years ago, orangutans that officially are now found only on Borneo and Sumatra also lived on mainland Asia. They were larger than today's orangutans and have been named *Pongo hooijeri*. These apes were closer in size to a gorilla. The man-sized Yeti could be a mainland orangutan or some related ape.

The smallest Yeti inhabits the warmer lower jungles and feeds on fruit and small animals such as frogs. It may be related to the orang-pedek of Sumatra that we will be examining later.

Startling new evidence for the Yeti has emerged recently. TV vet and naturalist Mark Evens, formally a Yeti sceptic, took an expedition into the mountains of Bhutan to make a documentary called *Lost Kingdom of the Yeti*. Water was taken from a pool in the mountains where the Yeti had been reported. From this, environmental DNA was taken. Known as eDNA for short, this consists of traces of DNA an organism leaves in the environment. It is a relatively new development and could prove an invaluable tool for cryptozoology as the techniques for extracting traces of eDNA improve. Back in the lab, the eDNA from the water was tested and several know species were discovered but there was also anomalous DNA. It came from a primate that shared 99 percent of its DNA with humans. Chimps share 98 percent. Whatever left that eDNA at the drinking hole was something unknown to science and closely related to man.

In China, a creature very much like the Yeti has been reported. The Yeren (wild man) is seen mainly in the Shennongjia Forestry District in north west Hubie Province. The forest covers 19,000 square miles. Records of the Yeren go back thousands of years in China. Pictograms of hairy humanoids have been uncovered that date to the Shang Period (1600–1046 BC). Hairy wildmen called Maomin are recorded in the fauna written about in the book *Classic of Mountains and Seas*.

The fourth-century book *In Search of the Supernatural,* a compilation of accounts of unexplained phenomena, records the story of a man named Tai Jing who encountered a six foot, six-inch-tall hair-covered, man-like beast in the mountains of Wuchang.

Li Shizen, the noted sixteenth-century polymath included the Yeren in his *Compendium of Meteria Medica.* Li classified it as a transitional species between ape and man.

In 1940, biologist Wang Xilin said that he examined the body of a dead female. The creature was six to six and a half feet tall. The eyes were very deep-set, the nose was huge and soft and the teeth and jaws enormous. The description in many ways, is similar to that of the pre-human Peking Man (*Homo erectus*), whose fossil remains have been found in the area.

In 1976, Chen Liansheng, a high-ranking local government official, had an encounter with a strange creature on a remote forest road in Shennongjia while travelling back from a work conference. It was 4 a.m. on May 4 and he had been asleep. He and his five colleagues were awakened by their driver shouting that something on the road was ahead of them. The driver was beeping his horn at a creature on the road that was blinking in the headlights.

> *The driver stopped about a metre away from the animal. Three of us in total got out of the car—it was so near that we could touch it. It was covered in red fur. Its face humanlike, with upright ears and a protruding mouth. Its eyes did not reflect light, like human eyes.*

One of Chen's colleagues threw a stone at the beast, but it leapt across a ditch and disappeared into the shrubbery.

The next day, Chen, who later became a Hubei Television journalist, filed a report about the incident with the Chinese Academy of Sciences and notified the local government. When he returned to the spot, he found around twenty bright red hairs, all of which were subsequently donated to museums or researchers, or lost.

The sighting prompted a government expedition the next year. It was led by paleoanthropologist Haug Wanbo and was composed of one hundred and thirteen members from scientific institutes including zoos, museums, and universities. Fifty-eight soldiers joined them. The expedition lasted a year and found huge manlike tracks that they cast and alleged hairs. Haung Wanbo concluded that the Yeren was a higher primate.

A number of recruits on the 1977 expedition went on to become solo researchers. Yuan Yuhao was one such man. When recruited for the expedition, he did not believe in the Yeren. However, in 1981 he saw the beast himself. It was about seven feet tall, its head was visible above the surrounding bamboo forest. He said...

> I've seen all of Shennongjia's wild animals, and there are none I don't recognise. Bears and golden snub-nosed monkeys cannot walk more than three to five steps on their hind legs.

Li Guohua, a lumberjack who has lived in the area since 1972, saw the beast in 1980 whilst on a solo expedition.

> I sat on a cliff to rest after hauling my equipment. Suddenly, I heard footsteps and saw a Yeren walking towards me. It was very tall, at least two and a half metres, taller than the tallest human. It was about sixty metres away. I was looking at the Yeren with all my concentration and it was likewise staring at me. I tried to target its lower body, since I wanted to injure it so it could not walk. But as soon as I pulled the trigger, I missed my target. I fired again and again, but did not hit it.

Li has no time for the armchair academics who reject the existence of the Yeren without ever bothering to search for it themselves.

> Academics have said that it was just a baseless rumours believed by common people who could not understand science, and their attitude has continued to this day. These so-called scientists who call themselves Yeren experts could sit at home drinking a single cup of tea for a lifetime and proclaim one sentence to deny the existence of Yeren: that they don't believe in it.

The Chinese Academy of Sciences organised another year-long expedition in 1980. They found droppings, tracks, hair samples, and nests. On December 18, 1980 two expedition members, Li Guohua and

Li Renrong, observed a long-haired primate sitting on a rock along the Xaiangshui River.

Hair samples of the Yeren have been tested with interesting results. Scientists at Fudan University found that the proportion of iron to zinc in samples of Yeren hair is fifty times greater than that of a human being and seven times greater than in other primates such as monkeys. This finding led to the conclusion that the hairs come from a higher form of primate unknown to science. The Fudan findings were independently confirmed by western scientists. Biologists at East China Normal University used a powerful electron scanning microscope to examine the hairs and came to the conclusion that the hairs were not human or any known primate. The Research Unit for Forensic Medicine of the Wuhan Hospital concluded after some specimen tests:

> We infer that the hair from these wild men could belong to an as yet unknown higher primate.

In the Garo Hills, in the state of Meghalaya, in north east India, a creature much like the larger type of Yeti is reported. Known as the *mande-barung* (forest man) it is said to resemble a huge, upright gorilla some ten feet tall and covered in black hair. Eleven-year-old Tengsim Marak saw the creature near the mouth of a cave in 2008. He said it was bigger than a man and had a face like a monkey. The creature was covered in black hair and seemed to be banging two rocks together.

In 2005, at a remote village called Rongri, the creature was said to have entered a hut occupied by a widow and her young child. The creature had stamped out the fire and sat down but had not harmed the woman, who was frozen with fear.

I have hunted the mande-barung myself and we will examine it further in the second volume of this work.

In Malaysia, the creature is called orang-mawas. In 2001, public relations consultant Eva Hawa was driving along Malaysia's main North-South highway when she saw the creature cross the road in front of her car.

It was hairy, it was big, it was about six to seven feet tall. He moved right across in front of my car. He had a hunch and walked like a very old man.

The year before, Liong Chong Shen, a farmer from Kampung Chennah, went outside to protect his durian saplings from what he thought were pigs. He noticed a strange odour and heard a grunting sound. He saw two orang-mawas about thirty feet from him. One was about six feet tall and covered in black hair. The second one was brown in colour and stood about five feet tall. They stared at him for a while then turned and walked back into the jungle.

In the jungles of Vietnam, the 'wildman' is called *nguoi rung*. American GIs during the Vietnam war who had run-ins with these creatures called them 'rock apes' because they hurled rocks at humans.

Veteran Kregg P. J. Jorgenson recorded a number of such stories in his 2001 book *Very Crazy GI: Strange but True Stories of the Vietnam War*.

In one case, a patrol in 1969 was caught off guard by enemy fire. The men hit the deck and lay in the vegetation. There, they heard the sound of something approaching. At first, they thought it was a man, but as it came into view, they saw it was a muscular, seven-foot-tall, ape-like creature covered in reddish hair. The creature moved fast but was winged by crossfire. The rock ape stumbled but did not fall. They enemy began to talk excitedly amongst themselves and seemed afraid. They fled, leaving a gun behind.

The soldiers moved into a more secure area but that night they were disturbed by screams and strange animalistic noises. On investigating the next day, they found a member of the Viet Cong who had been torn apart. They were so disturbed they moved out of the area.

Ornithologist Professor Vo Ouv took an expedition, in the early 1970s to Kontum province near the boarders of Cambodia and Laos. He found and cast a print wider than a man's and too large for any known primate in the country.

We must remember that several large mammals were found in Vietnam in the 1990s.

Ngo Hoang, who between 1950 to 1952 was an armed agent of the Ministry of Propaganda in the hostile back country of Dac Lac and later was a member of an MIA committee in the years 1988 to 1991. As a result, he was well acquainted with all the forest areas of Tay Nguyen. He relates the following story…

We were able to discover these forest men in about 1950 in an area in the vicinity of the Chu Bia mountain chain (which now belongs to the district of Dak Nong—Dak Lak). The footprint this forest man left behind was one and a half times as big as that of a normal man (length: about thirty cm, breadth about twenty cm). The big toe is slightly separated from the others. In the middle of the sole of the foot are many folds. After that, some of my comrades of a Pioneer unit saw a big 'man' entirely covered with grey hair. They thought it was an orangutan and wanted to kill him. Fortunately, at that time we had strict orders not to shoot. That is why this man could escape. My comrades told me that he ran very fast, as fast as only a forest creature can run, but no man can run…

In Pakistan, in the cool, high mountains on the borders of Afghanistan, the hill tribes speak of the barmanou (strong one), a muscular, hairy manlike beast encountered by shepherds, hunters, and honey farmers who live and work in remote areas.

Spanish zoologist Jordi Magraner investigated the creature from 1987 until 2002 when he was murdered. The case and its motives are still unknown but Magraner was a vociferous campaigner for the rights of the non-Islamic hill tribes. He collected more than fifty first-hand accounts of the creature.

Lal Khan, a fifty-five-year-old Gujar shepherd, saw a barmanou in 1987. He was leading his goats to pasture. At around 1:00 in the afternoon, he was herding them across a slope opening on a prairie at 11,300 feet of elevation. The noise of the goats woke up a barmanou lying at the foot of a large fir tree. The lower branches of the tree, growing in front of cliff, created a shelter reaching ground level. The shepherd saw the creature coming out of its shelter. It was about thirty feet away from him. The creature was robust and completely hairy over its whole body except on its cheeks, the palm of its hands, its knees, and ears. It had a beard, short head hair and broad shoulders, massive teeth, without fangs. Its hair was dark, except on its chest where it was white.

It picked up a rock and hurled it at the shepherd. It then turned around and calmly crossed the prairie using a stick in its left hand in support. It moved uphill and disappeared behind a ridge. Now and then, the wild man stopped to gather plants and to eat them. It took a long detour along elevation contours. The sighting lasted more than ten minutes. The Lal Khan, frightened, gathered his goats and went back down the hill.

On a day in April 1987, around noon, Nur Hamid, another fifty-five-year-old Gujar shepherd, was gathering morels in the mountains with his sister in a conifer forest at about 9,800 feet. Suddenly, they saw a barmanou. It was about sixty-five feet from them, downhill, crouching in a bush in the fetal position. When it saw them, it approached. He and his sister threw rocks at it and started shouting. A man came to help. Finally, the creature was hit on the head with a stone and ran away fast. It disappeared in the forest, downhill. It was about five foot seven or nine inches tall. Its dark skin was entirely covered with reddish hair. Its long wild head hair came down to its shoulders. Its build was muscular.

To the north in Tajikistan, a similar creature is called the Gul. Journalist Ben Judah collected stories from the Romit valley when he visited it in 2010. One peasant man he and his guide gave a lift to said he was attacked by a creature with long dangling breasts and a covering of black hair ten years before. He had been gathering firewood when he saw a figure he thought was human. Shouting out to it, he attracted its attention and it charged at him making a *raggh* sound. The man fled.

Another witness, fifteen-year-old Assodin, had seen a Gul just a few days before whilst gathering wild onions in the mountains. He heard its whistling call, then saw it clambering over rocks. He likened it to a monkey, a creature that is not found in Tajikistan.

Ben noted that the stories were only present in areas with forest. Above the tree line, the stories petered out.

Boris Porshnev collected stories from Tajikistan as well. Abdulhamid Abdurakhmanov, sixty-seven, recorded the story of his grandfather, a hunter named Allaëra. It was in Jirgatal—east of the Karateghin Range. Five hunters went up into the mountains for wild sheep and spent the night in the woods. In the middle of the night, the hunters woke up because their dogs were barking. The threw dry grass on the fire to light up the area. They saw at close range a Gul. It was a huge manlike creature almost six feet in height. It was all covered with hair. It gave off a very strong and unpleasant odour. The creature stood for a while and looked at the hunters, and then the wind drove the smoke and sparks into his face. Then he was gone, and they could hear how he broke branches and rustled bushes in the forest. In the morning, the hunters found and measured his tracks.

Abdugaffor Saburov, forty-six years old, a disabled World War II veteran, was hunting in the area in the fall of 1948 around Romit, near the village of Canas. At a distance of 1,640 to 1,960 feet he saw a huge humanoid creature, covered with hair, with large hands and feet. The hair was longest on his shoulders, thighs, and breasts. It gave off a powerful stench.

In the former USSR the most famous of the wildmen was known as the Almasty. These manlike beasts were reported from the Caucasus, Altai, and Pamir Mountains. Ranging from human-sized to around seven and a half feet tall, the Almasty is covered with hair except for the face. It has a thick brow ridge, sloping brow, flat, wide nose, a wide mouth with thin lips and a muscular build. It does not use fire but can hurl rocks and swing clubs.

The most detailed account comes from Major General Mikhail Topilsky of the Red Russian army, who was trailing a troop of defeated White Russians through the Pamir Mountains in 1925.

> In the autumn of 1925, together with a scouting party, we were engaged in tracking down a gang of anti-Soviet guerrillas which was operating in the Western Pamirs. They were trying to shake us off by going to the Sinkiang via the Eastern Pamirs. On our way through the highland villages in the Vanch district we had heard stories about hairy man-beasts, monstrous creatures (I don't remember the local name for them) that lived in the mountains. They were said to be hostile to humans; although they didn't usually

attack first. (...) Once when we were following the gang's tracks along a mountain path and had already reached the permanent snowline, we saw some tracks running across the path. Our dog took up the scent but refused to follow the tracks. They were very clear and there could be no doubt they were the prints of bare human feet. They continued for some one hundred and fifty metres and stopped at the foot of a sheer, barren cliff, which a man could hardly have climbed. Our doctor studied the tracks thoroughly and decided that they were human footprints beyond all doubt. (...) Continuing our chase, we caught up with what was left of the exhausted gang, which had stopped for a rest at a place where a glacier was split apart by a stone cliff. The upper tongue of the glacier hung from the cliff, in which there was a crevice or cave. We surrounded the gang and took up a position above where they were resting. A machine-gun was placed in position. When we threw the first grenade, a man (a Russian officer) ran out onto the glacier and started shouting that the shooting would make the ice cave in and that everyone would be buried. When we demanded that they surrender he asked for time to talk it over with the other guerrillas, and went back into the cave. Soon after, we heard an ominous hissing as the ice began to move. At almost the same moment, we heard shots, and not knowing what they meant decided that it was the beginning of an assault.

Pieces of ice and snow started falling down from the cliff, gradually burying the entrance to the cave. When it was almost buried, three men managed to escape, and the rest (we learned later that there were five) were buried under the debris. Our shots killed two of the guerrillas and seriously wounded the third. (...)

We questioned him and he gave us the following information. While they were discussing our order to surrender, some hairy, manlike creatures, howling inarticulately, appeared in the cave through a crevice (which possibly led upwards from the cave). There were several of them, and they had staves in their hands.

The men tried to shoot their way through. One of the guerrillas was clubbed to death by the creatures. Our translator received a blow from a staff on his left shoulder as he rushed to the cave entrance with one of the monsters hard on his heels. It ran out of the cave after him, but was shot and buried under a snowslide.

To check up on this strange story we made him show us the exact spot and cleared the snow away. We recovered the body all right. It had been shot three times. Not far off we found a stick made of very hard wood. At first glance I thought the body was that of an ape: it was covered all over with hair. But I knew there were no apes in the Pamirs. Also, the body itself looked very much like that of a man. We tried pulling the hair, to see if it was just a hide used for disguise, but found that it was the creature's own natural hair. We turned the body over several times onto its back and its front, and measured it. Our doctor (who was killed later that year) made a long and thorough inspection of the body, and it was clear that it was not a human being.

The body belonged to a male creature one hundred and sixty-five to one hundred and seventy centimetres tall, elderly or even old, judging by the greyish colour of the hair in several places. The chest was covered with brownish hair and the belly with greyish hair. The hair was longer but sparser on the chest and close-cropped and thick on the belly. In general, the hair was very thick, without any underfur. There was least hair on the buttocks, from which fact our doctor deduced that the creature sat like a human being. There was most hair on the hips. The knees were completely devoid of hair and had callous growths on them. The whole foot including the sole was quite hairless, and was covered by hard brown skin. The shoulders and arms were also covered with hair which got thinner near the hands, and the palms had none at all, but only callous skin. The colour of the face was dark, and the creature had neither beard nor moustache. The back of the head was covered by thick, matted hair. The dead creature lay with its eyes open and its teeth bared. The eyes were dark, and the teeth were large and even and shaped like human teeth. The forehead was slanting and eyebrows were very powerful. The prominent cheekbones made the face resemble the Mongol type of face. The nose was flat, with a deeply sunk bridge. The ears were hairless and looked a little more pointed than a human's, with a longer lobe. The lower jaw was very massive.

The creature had a very powerful broad chest and well-developed muscles. We didn't find any important anatomical differences between it and man. The genitalia were like those of a man. The arms were of normal length, the hands slightly wider and the feet much wider and shorter than a human.

We did not know exactly where we were, because no accurate maps of the Pamirs were then in existence. But we must have been somewhere between the Yazgulem and the Rushan Ranges. As we had completed our task we had to return. (...) The nature of the dead creature presented us with a problem. But it was impossible to take the body with us on the difficult trek that lay ahead. Also, it could have caused complications with the local population. We could say, of course, that we were carrying the body of an animal, but the creature looked too much like a human being. We thought about skinning it, but it was too much like skinning a man. In the end we decided to bury the creature where we had found it. We did not try to enter the cave because we were afraid of another cave-in.

Jeanne-Marie Koffman explored the Caucasus extensively, and during her travels met, Talib Kumyshev, sixty-seven, a Kahardian, who was a respected elder man of the village of Kamennomos. His testimony is highly revealing:

It was probably in 1930, or 1931, or 1932, in June or at the end of May, when our cattle left for the alpine pastures of Elbrus.

I was chief of the group. We had left to inspect the herds with the veterinarian.

Well, rain had surprised one of my shepherds, Shaghir Zagureyev, very high up on the slopes, and he had gone to take refuge under a rocky overhang. As he approached it, he saw there were three Almastys sitting under it. Shaghir was a little frightened, but as the rain was by then falling much harder, he decided to stay there anyway, though at a distance from them. They looked at one another. Then, the rain stopped and Shaghir came down to the farm. He did not say anything to anyone.

Very early in the morning, I was awakened by cries, a tremendous noise, and I saw that the shepherds were running to assemble their herds and were taking the cattle down the valley. 'Why are they leaving?' I asked. 'There are Almastys under the rock, up there.' At that moment Shaghir declared, 'It's true, there are three Almastys sitting up there, I saw them yesterday evening.' I was then really angry. I said to Shaghir, 'You're an idiot. You were frightened by a bush.'

'No,' said Shaghir. 'I saw them.'

'Well, why didn't you tell anyone?'

'Because the old people have warned: when you see an Almasty for the first time, if you tell anyone about it, you'll get a bad headache. Well, for me, it was the first time that I have seen one.'

I continued not to believe all this. They said to me, 'Okay, go ahead, go see for yourself.'

We were about ten to fifteen people making a half-circle around that rock. We stayed there until dinnertime. Some went away, and others came up. Three Almastys were seated under the overhang, two of medium size, and the other bigger. The one which was the biggest was in the middle. They were sitting on rocks, facing us,

160

hunched over, with their heads down. From time to time they raised their heads slightly, and looked at us from under their brows.

Their heads are very ugly, not nice at all. Their faces resemble human faces, but the nose is shorter and flattened. The eyes are slanted and reddish. The cheeks are very prominent, like those of a Mongol or a Korean, but more so. The lips are thin. The lower jaw is receding, as though cut on a bias. The hair is long, like that of a woman, and tangled. The entire body is covered with shaggy hair, resembling that of the buffalo. In some places this is long (torso, chest) and in other places it is shorter (arms, legs).

The big one had the chest of a man. The others had the breasts of a woman, but extremely long and covered with hair. The hair was very dirty. The stink was so strong that we could not stand it. The odour was like that of wild flax, when it grows thickly. Once, the one seated on the right mumbled something. I did not see their hands clearly, as they were held between their legs. The legs are rather short and bowed. The foot is like that of a man, but more spread out. All were wearing, wrapped around their waists, an old piece of a shepherd's cape. A young shepherd proposed to throw a lasso around one of them and bring it into the village. But all the others cried out that it is forbidden, that they must not be harmed, and that they must not be disturbed. I watched them from a distance of three or four metres, and I even approached to within about one metre. Did I touch them? I should say not! If you touch them, as Allah is my witness, you could no longer eat with your hands afterward, they are so dirty, stinking and repulsive. I remained one and a half to two hours. When I left, other shepherds were arriving. I have heard my father recount that they suckle on cows.

Reports of the Almasty have been made close to the borders of Europe. One of the most detailed, multi-witness cases happened in 1987 on the Kola Peninsula.

In the summer of 1987, a group of six youths aged fifteen to eighteen from the town of Lovozero fished and picked berries and mushrooms around a lake of the same name on the Kola Peninsula in western Russia, around 100 miles from the Finnish border. The boys had built a wooden cabin. The building was raised up on fir stumps as the area was prone to flooding in spring. On August 11, the boys were sitting around a campfire. They had the odd feeling they were being watched. Five of them retired to the shack. One of them, Sasha Prikhodchenko, lay down to sleep by the campfire. As he peered out from under his blanket, he saw a pair of huge, hair-covered legs approaching the shack from behind. The legs were visible due to the shack being raised on tree stumps. Sasha ran back to the shack and told his friends. Looking out they saw a huge, manlike creature covered in grey hair. It circled the shack. Terrified, the boys used a stick to bolt the door and stayed awake all night. They nicknamed the monster 'Afonya'.

In the light of day, the boys became braver. Nothing happened all day but as night fell, they decided to scare Afonya off with noise. They played pop music full volume on a tape recorder and threw rocks into the bushes. As they sat around the fire a rock flew out of the bushes and landed in the fire. Shortly after another struck the cabin. The group took shelter in the cabin as it was bombarded with rocks. One of them looked out of the door whilst protecting his head with a metal pot. He saw the Almasty moving back into the forest. Later Afonya returned and beat the outside of the cabin. The boys reasoned that the creature had been sleeping in the cabin in their absence and was now trying to drive them out.

Next day, they secured the door with a stick and returned to the town on their motorboat. They told other people of their adventure but were roundly disbelieved and laughed at. They returned to the cabin and found the stick still in place. That night Afonya failed to appear. One boy suggested that they tried to surround the Almasty and kill it with axes but the others feared that the monster would kill them all. Instead they decided to try and befriend the beast.

The following day, they left a table of food out for Afonya and went back to their boat. Later, they saw the creature on the shore. He seemed alarmed by their outboard motors and vanished. The boys returned and found the food untouched. They decided to stay in the cabin again.

Before going to sleep, one of the group, Slava Kovalev, stepped outside of the cabin to urinate. As he finished, he looked up to see the Almasty only six feet from him. Slava ran into the cabin and bolted the door with a stick. Afonya came up to the door and pushed it, snapping the stick like a matchstick. As the door swung open, the monster stood in the doorway. The panicking boys scrambled under the beds but Afonya did not enter but simply banged on the door. The frightened boys dare not approach the door to close it. They stayed awake all night and heard the Almasty leap onto the roof and walk about. As the sun rose, he left.

The following day, three more boys arrived from the town and the original six were emboldened to stay. As they sat around the fire that night, Afonya appeared. He was squatting in an ape-like stance on all fours some forty-five feet away. He bounded back and forth on all fours then charged. The boys scrambled into the shack. Slava held the door and shouted for help. Roman Leonov tried to push an axe handle through the metal hoops that had held the now smashed stick bolt. As he did so the monster gripped the half-opened door. Roman got a close look at Afonya's hand which he described as like a man's only much bigger. It was sparsely covered in grey hair and had dark skin.

The boys frantically pulled at the door as they tried to slide the axe handle through the bolts. The Almasty pulled back and ripped the door open. Roman got another good look at the beast, this time his face.

I saw his face right in front of mine. Mind you, he was standing on the ground and I was on the floor which is raised high above the ground. His face is brown and wrinkled. Somehow, I didn't notice his nose and mouth. I just remember the eyes glittering, angry and reddish. They are set far apart, like a horse's, sort of looking sideways. The forehead is wide and polished, the head is round like a ball. I can't remember how I got under the plank-bed and dropped the axe. I came back to my senses because of a strong knocking on the walls and the boys yelling like mad. The knock was so strong that I dashed back to the door. It was open

*and I saw Slava Kovalev by the fire and moaning from pain. As
I learned later, when Afonya pulled the door sharply, Slava was
thrown out of the cabin and hit his shoulder against the door-
frame. He was so angry when he returned to the cabin, he said
he would go out and face our adversary, but we asked him to
stay inside the cabin.*

Afonya leapt onto the roof again and stomped around so loudly the boys
feared he would crash through the roof. The monster apparently burnt
his hand on the hot chimney that was attached to the burning stove. The
Almasty made a mooing sound, jumped down and retreated into the forest.

In the morning, six boys returned to town and three remained at the
cabin. They spent the day fishing. On returning to the cabin, they saw
Afonya approaching quickly. The boys climbed onto the roof via a ladder
and cringed in fear but the monster simply turned and went back into the
forest. As soon as he had gone, the boys ran for their boat and motored
back to the town.

Back in town, the boy's reports were treated with scorn once more.
They were even accused of taking drugs.

On August 18, the group returned with three more boys from the town.
Now twelve boys were searching for the monster. After seeing Afonya
prowling the shore from their boats, the three newcomers chickened out
and turned their boat around. The Almasty hurled rocks at the boats as
well as a branch from a birch tree that almost struck one of the boats.

They returned the next day and fished from their shore except for
Sasha Prikhodchenko who was cleaning the table outside of the cabin.
The boys in the boats saw Afonya approaching and shouted a warning
to Sasha who thought they were joking as on all previous occasions the
creature had only appeared at night. But looking up he saw the creature
only fifteen feet from him. Sasha ran for the cabin and bolted the door
with an axe handle. The others ran for their boats.

The boys returned later to check on Sasha and saw the Almasty on
the shore. He kept pace with their boats and did not let them come to
shore. When he eventually vanished into the trees, they cut the motors
and rowed to shore. Three boys, Slava Kovalev, Sasha Sveilis, and Roman

Leonov, decided to land. As they walked towards the cabin, Afonya leapt from the bushes and chased them. Slava and Roman jumped into their boat and started the motor. Sasha's boat was stuck and, as he struggled with it, the motor fell into the water. Afonya came up behind him and the other boys shouted to him to jump into their boat. Sasha, however, was paralysed with fear.

The Almasty crouched, then stood, watching the boy. The others rowed alongside and told Sasha to climb into their boat. As he tried to do so, he glanced back at the monster and fainted. The other boys pulled him into their boat and splashed water on him to awaken him. Rowing out again, they waited till the beast had retreated.

Sasha Prikhodchenko, who had been watching from the cabin, shouted out that the creature had retreated beyond a creek. Arming themselves with axes, rocks, and hammers, the others returned to try and rescue Sasha from the cabin.

Afonya ran out of the forest again, right into the group, cutting it in two. One of them, Slava Surodin, hurled a rock at him. It struck Afonya in the shoulder and he made a mooing sound. Two of the group ran for the cabin, the others for the boats. The Almasty chased the boys who ran to the cabin.

One of them, Ivan Dyba said...

As I felt him literally breathing down the back of my neck, my legs failed me. I snatched at Zhenya Trofimov running in front of me and we both fell to the ground. We crawled and scrambled into the cabin.

The terrified boys hid inside as the monster pounded the walls and peered through the window. Yet again the Almasty retreated into the forest and the other boys tried to rescue their comrades. As if he were playing cat and mouse with them, Afonya charged out of the undergrowth again. Sasha Prikhodchenko and Zhenya Trofimov ran back to the cabin whilst Ivan Dyba leapt into the water, fully clothed, and swam to a boat. Those on the boat pulled him aboard and motored to the opposite shore to dry his clothes.

Deciding to alert the authorities, the three boys on the boat returned to town. Arriving at 2 a.m., they went to the town's executive committee. The woman on duty called a patrolling militiaman. After listening to their story, he called a senior game warden called Kuznetsov who arrived an hour later. Laughing at the boy's story, he said...

I am not going to alert anyone in the dead of night. I don't want to become a laughing stock on account of your 'snowman'. I'll send someone over in the daytime.

On August 20, help came to the boys trapped in the cabin in the form of a game warden called Igor Pavlov and his two assistants. All had rifles. After listening to the story, they arrived at 11 p.m. The summer that far north still giving plenty of light, the boys alerted the men when Afonya emerged. Igor had been told by his superiors that the boys had been frightened by an 'old bear' but what he saw some eighty-two feet away looked more manlike but Pavlov saw it was clearly not human. The monster ran back into the forest but the warden was able to gauge the creature's height against a branch sticking out from a tree. Afonya was at least eight feet tall. He noted that the Almasty's arms reached its knees and it was covered in grey hair. It ran with long strides.

The adults and boys stayed in the cabin that night. The building was bombarded with rocks. In the morning, the boys left home but the wardens stayed on. They found a number of tracks imprinted more deeply than an adult man. They spent another night at the cabin but were not disturbed.

The next day, several adults accompanied the boys to the site with torches and cameras. Afonya hurled rocks at the cabin but did not show himself. Soon after the area was overrun with film crews and reporters. Afonya by this time had vanished.

The location of these events is close to the border with Finland. Scandinavia is well known for its tradition of trolls, shambling, hairy, club-wielding, man-eating giants of ancient legend. Lars Thomas of Copenhagen University was studying ancient texts pertaining to a legendary Danish king who loved to hunt. His favourite quarry were trolls because of their savagery when cornered. The descriptions of trolls in these writings spoke of tall, muscular, hair-covered creatures much like men. They had thick brow ridges, long arms, deep-set eyes, and the females had long breasts. Trolls had no fire but could hurl large rocks and use clubs. The description matched up uncannily with modern day descriptions of the Almasty that I had recently collected from witnesses in the Caucasus Mountains in Russia. It is likely that these creatures, on occasion, cross over into Europe.

Though now uncommon, sightings of trolls persist in remote areas of Scandinavia. Lars Thomas has collected a number of them. Some of the most spectacular come from Sweden. Most reports have been in the vicinity of Kebnekaise, the highest mountain in Sweden.

In 1953, a mountaineer described seeing a group of creatures that looked like white gorillas on the mountain. The creatures vanished when he tried to get closer.

The Sami people call trolls '*Stallo*' and it was a student of Sami heritage that saw one of the beasts in 1961. He was walking to Kebnekaise from Nikaluokta when he saw a large white figure running across the bottom of a valley some nine hundred and eighty feet away. It stood on two legs and looked like a large man covered in shaggy white fur like an 'unwashed polar bear'.

A Sami woman and her two young daughters were driving towards Nikaluokta along the northern edge of Laukkujarvi Lake. The woman had slowed the car down as her youngest daughter had complained of feeling car sick. Suddenly, she had to slam on the breaks as a huge creature appeared by the side of the road. It was looking directly at the car and her daughters began to cry. She described it as bigger than the biggest man and covered in dirty white fur. It had yellow eyes. The foul-smelling beast grimaced at them then crossed the road no more than thirty-three feet from her car. She said the stench was like a dead reindeer that had been lying in the summer sun.

In the summer of 1984, two Danish birdwatchers were visiting Abisko National Park, north of Kebnekaise. They had, for several days, been walking around Abiskojaure Lake and had seen a brown bear. One afternoon they had set up a picnic near an area with thick underbrush. Hearing growling from the thicket, they thought it was a bear and walked slowly away, leaving the food. Looking back, they saw a manlike creature covered in white hair eating the food.

A friend of Lars Thomas' father told him of how, as a youth, he had fought against the Red Army in the forests of Eastern Finland. Their sky patrols were constantly watched by a race of hairy wildmen they called *hiisi*. The creatures would only vanish if the soldiers were fighting the Red Army. The soldiers once came upon a wounded hissi that had been shot during one of these conflicts. It looked like a powerfully built, hair-covered man. They had their own wounded to attend to, so they left the creature where it was.

Lars' grandfather also had a Finnish friend who fought against the Red Army. He too saw a hiisi when he was in Finland. When he was on guard duty one night, he saw a hiisi only sixty-five feet from him standing against a tree. He felt no hostility from the creature that looked like a large, broad-shouldered, hairy man. The creature was carrying a long stick, possibly a spear.

In 1997, a group of moose hunters described being watched by a hairy man or ape-like beast. One witness, a lawyer from Helsinki, said the creature was leaning against a long stick that may have been a spear.

Three birdwatchers, two Danes and a Swede, were bird watching in Finland in 2004. They were looking at a great grey owl in a tree when a hooting sound, unlike that of an owl, scared the bird away. Soon after they saw two humanoid figures covered in dark hair running fast through the forest.

Professor Yenshööbü ovogt Byambyn Rinchen investigated accounts from Mongolia, where the creature is known as the almas. One story came to him from a Mr. Damdin of the State Museum of Ulan Bator. He had spent several months in the Khovd and Bayanolgy provinces, and wrote:

> It happened at about ten o'clock of 26th June, 1953. I remembered the time, day, and month because this day had utterly surprised me and was engraved on my heart. At dawn of that day I went to search my lost camels in the direction of the so-called Red Mountain of the Almases. It was a beautiful, sunny morning when I dropped into the ravine. The wind spread a fragrance of

highland flowers and herbs but I was in a hurry to leave before a midday heat in this labyrinth of canyons and ravines. My camel climbed up and down in the craggy defile. Suddenly I saw in the corner of secluded ravine under two small ammodendron bushes something of camel-colour.

I approached and saw a hairy corpse of a robust humanlike creature dried and half buried by sand. I had never seen such humanlike being covered by camel-colour brownish-yellow short hairs and I recoiled, although in my native land of Sinking I had seen many dead men killed in battle. But who was this strange dead thing—man or beast? I decided to return back and thoroughly examine it. I approached once more and looked down from my camel. The dead thing was not a bear or an ape and at the same time it was not a man like Mongol, Kazakh, Chinese, or Russian. The hairs of its head were longer than its body. The skin on its groin and armpits was darkened and shrivelled like the hide of a dead camel. I have also examined a terrain near its body and never found any rests or wears. Fear seized my heart. I remembered the old tales of Vetala-Vampires and thought I was to see before me one of them. And I hurried away. After my return home, I had informed our local administration and Mr. Chimeddorje, manager of Fruit Growing Station, but no one gave attention to my account.

According to Professor Rinchen, a Mongolian named Anuhu and a companion were driving camels in the Southern Gobi when they noticed a thickset shaggy biped running away from them. Both decided to try and catch the runaway on camels. But when the fast Gobi camels caught up with the creature, they saw it was covered with short hair. The beast then issued a sharp cry, that frightened and stopped the camels. And then the *almaska*, as they are called in that area, ran back to the stone ridge, and scrambled up it with the help of arms and legs, and disappeared.

An account of my own investigations into the Almasty in Russia can be read in the second volume of this book.

These wildmen of the former USSR, Mongolia, and Central Asia sound much more man-like than the hulking Yetis of Tibet, the Himalayas, and northern India. The Russians took them so seriously that they even had on official Snowman Commission to investigate the creatures. At the time, it was thought that they may be a relic form of Neanderthal. Since then, however, we have discovered that Neanderthals looked very much like us. It has been said that if you washed and shaved a Neanderthal and put him in modern clothes, he could walk down the street in any major city without raising anyone's eyebrows. Sure, he may look somewhat ugly by our standards, but he would clearly be human. Neanderthals used fire, made sophisticated tools and clothes, and may have even had a concept of religion and afterlife. They sometimes buried their dead with grave goods. This is clearly not what we are dealing with here.

It is more likely that the wildmen reported today are an offshoot of a much more primitive species of hominin. In recent years, both fossil, sub-fossil, and genetic evidence has unearthed many new species of human relatives. We know that Neanderthals interbred with modern man. The genomes of all non-African people contain 1.5–4 percent Neanderthal DNA.

In March of 2010, a tiny fragment of finger bone was found in the Denisovan Cave in the Altai Mountains. The bone was so well preserved that the whole genome was intact within it. It turned out to be from a new species of archaic humans that have since been named the Denisovans. Only fragments of this species have been discovered—the finger bone, a toe bone, and two teeth—so the appearance of the Denisovans is unknown. However, they interbred with both Neanderthals and modern man. Between 4 and 6 percent of the genome of Melanesians (people from New Guinea and the surrounding islands) is inherited from the Denisovans.

Some hominins are suggested not by any fossil remains but by genetic markers on populations of modern man. As wells as Denisovan DNA, Melanesians appear to have inherited DNA from another hominin currently unknown from the fossil record. Yet another unknown archaic hominin appears to have left genetic material in the populations of sub-Saharan Africans.

In 2003, some sub-fossil remains were found in the Liang Bua cave on the Island of Flores in Indonesia. The remains were of a tiny species of hominin that were named *Homo floresiensis*. The remains were dated to around fifty thousand years ago. The creatures would have stood no more than three feet seven inches tall. The remains were found with stone tools and weapons as well as evidence of fire making. They seemed to have hunted giant rodents and pygmy elephants that lived on the island.

Homo floresiensis was thought to be a dwarf island form of *Homo erectus*, the hominin that was the ancestor of not only modern man but Neanderthals, Denisovans, *Homo heidelbergensis*, and *Homo antecessor*. More recent examination of the remains, however, showed that they were more closely related to *Homo habilis*, a more primitive hominin that has never been recorded outside of Africa and died out some one and a half million years ago. So, *Homo floresiensis* was not only half a world away from where it should have been but also nearly one and a half million years out of time. It also begs the question 'what else is out there?'

More recently, two as yet unnamed species of hominin dating to only ten thousand years ago (an eye blink in evolutionary terms) have been unearthed at Red Deer Cave in south west China. They seem also to have affinities to *Homo habilis*. It seems that *H. habilis* may have had its own lineage outside of Africa alongside of *H. erectus*.

Back in Flores, *H. floresiensis* has been linked to the legend of the ebu gogo. The Nage people of Flores speak of the *ebu gogo* or ancestor that eats everything. These were upright, walking, hairy creatures with wide and flat noses, broad faces, and large mouths. They spoke in a mumbling language and were infamous for stealing the crops of the Nage and killing their livestock with bamboo spears. It is said that the Nage retaliated by inviting the ebu gogo to a great feast and got them drunk on palm wine. When the creatures returned to their mountain cave, the Nage blocked up the cave entrances with dried palm leaves then set them alight, smoking the ebu gogo to death. A handful survived and still live in the jungles of Flores.

Some think there may be a grain of truth in the story and link it to a volcanic eruption in 1850, which is around when the story is set. A localized eruption could have wiped out a colony of the creatures. A near-identical story is found in the folklore of the Vedda people of Sri Lanka. Known here as the Nittaewo, the three-foot-tall, hairy, nuisance-making creatures were also trapped in a cave by the Vedda and smoked to death.

British primatologist, W.C. Osman Hill, led an expedition into the region in 1945 and found widespread belief in the Nittaewo still being alive on Sri Lanka. He thought that they could be a surviving form of *Homo erectus*. Captain A.T. Rambukwella thought that the Nittaewo may have been a strain of surviving Australopithecines. He led an expedition to the Mahalenama area in search of the Nittaewo in May 1963. Of course, neither of these men knew of *H. floresiensis* who makes a better model for the Nittaewo.

Dr Salvador Martinez, a Spanish anthropologist, claimed to have seen a Nittaewo in Sri Lanka in 1984. It was manlike but covered in hair. It made strange sounds before fleeing into the forest.

On Flores, too, it is said that the creatures still exist. Anthropologist Gregory Forth has interviewed some witnesses who may have encountered live ebu gogo. One such man was Jon Lulukumi, a civil servant and former headmaster of a school. In 1975, when he was in his early twenties, he and a friend were hiking round the Ebulobo volcano. Late one afternoon, they set out from the village of Wudu, east of Ua and headed towards the island's south coast. In the light of a full moon in forest interspersed with gardens (semi-cultivated areas on the edge of the jungle), they saw a strange figure emerge from behind a banana tree trunk. At first, they thought it was a little old man. The figure, who was about one hundred feet away, was about three foot nine to four foot two tall. It had a broad chest and long, dishevelled, black hair on its head. It was naked and had a dark complexion. The creature ran away when it noticed the youths.

Another sighting was reported by Yohanes Soda Ule, an elder of the Nage people. One day in 1965, he was approaching a stream in a forested area some five kilometres from his village. On the other side of the stream, from a distance of thirty-three feet, he saw a humanlike figure running away from him. It was five feet tall and had dark hair on its head. The body was copper or rusty-coloured. It had long arms that extended to the knees and upper legs and torso that were comparatively short. He found tracks that had a narrow heel and wider front part of the foot sporting large toes.

A third witness was Mikel Goa Na'u, the leader of a prominent clan called the Mudi. He claimed to have seen a group of creatures entering a cave near Hobo Bo. The creatures were walking upright, naked, and less than three feet tall. They had head hair and genital hair like humans, but their faces were somewhere between human and monkey in appearance.

There is a distinct possibility that *H. floresiensis* may still be with us today.

Before we move on, these newfound or newly suggested hominins may have one last surprise for us, back in the formerly named USSR.

In 1850, a female Almasty was captured near the isolated mountain village of T'khina in Abkhazia. She was six feet nine inches tall and covered with reddish hair. Her skin was black and she had a flat, ape-like nose, sloping brow, and powerful jaws. She was taken to the farm of Edgi Genaba and was at first savage. She was kept in a stockade. After a while she became tame and was named her Zana, meaning dark. Zana never learned to speak but would do menial chores around the farm such as carrying bags of flour. She was immensely strong. Zana disliked being indoors and slept outside in a hollow she dug for herself. When not working she would often bang rocks together.

Zana had a weakness for alcohol and if she got a hold of it would drink until insensible. In this condition she was raped by a number of the village men. If the reader finds this hard to credit, then recall that in 2003 a female orangutan called Pony was doped, shaved, chained, and used for sex by men in the village of Kareng Pangi, Central Kalimantan, Borneo before being rescued and rehabilitated. Zana became the mother of a number of hybrid children. The early ones she tried to wash in a river, but the cold water killed them. Subsequently, other offspring were taken away and raised by the villagers. The children grew into normal members of the community and married in later life. They were swarthier than the local people and far stronger. They could crack nuts with their jaws and lift a man sitting on a chair with their jaws alone.

Zana died in about 1890 but her youngest son, Kwhit, died in the 1950s. His skull was unearthed in the 1970s and kept at the Darwin Museum in Moscow until it was purchased by Russian cryptozoologist Igor Burtsev. The skull had some archaic features such as a very robust jaw and large size.

Over in England, Bryan Sykes, Professor of Human Genetics of Oxford University, had begun an interesting project. Together with Michel Sartori, the Director of the Museum of Zoology in Lausanne, Switzerland, Professor Sykes instigated The Oxford-Lausanne Collateral Hominid Project. The idea was to bring hard science into the search for manlike monsters. Sykes and Satori invited people to send them supposed hair samples from unknown primates such as the Yeti, Sasquatch, Almasty, and orang-pendek.

Sykes has researched human origins for over twenty-five years via the study of mitochondrial DNA. This is inherited from the maternal line and is generally the best-preserved DNA. Mitochondria are found between the cell wall and nucleus of each cell and release energy, ergo they are relatively abundant. Simply put, the professor has perfected a technique that examines a DNA segment called 12S RNA, part of a gene that helps mitochondria assemble the enzymes required for aerobic metabolism. This sequence is known for all established species of mammal. Hence there could be no confusion in any sample sent to the Project. They would be from one of the known species or from something new. A hypothetical new species could have its place on the genetic tree revealed by its closeness to other species. This meant that human contamination could be avoided. Even Neanderthal 12S RNA differs from modern man.

All the hair samples sent in turned out to be from known species. However, Sykes secured a tooth from the skull of Kwhit. Extracting mitochondrial DNA from the tooth Professor Sykes found that it was 100 percent sub-Saharan African. This was confusing as Zana was clearly no kind of modern human. Her behaviour and appearance seemed to be far more primitive than a Neanderthal. Further work showed the DNA was from an exceptionally ancient lineage from Western Africa and furthermore it may have been pre-Homo-sapien. Sykes thinks that this lineage may have left Africa over one hundred and fifty thousand years ago, before modern man. If he is correct, then Zana may have been an unknown species of pre-human hominin, a species still lurking in the Caucasus and other areas today.

I was in contact via email with Professor Sykes. Like all good scientists he is proceeding with caution. He and another geneticist are still working on the sample. He has likened Zana's DNA to fragments of old photographs that have filtered down through time. I await the results with bated breath.

It is on the Indonesian island of Sumatra that the most consistent and convincing reports are found. Orang-pendek means 'short man' in Indonesian. The creature is said to be powerfully built and immensely strong but relatively short at around three to five feet in height. It walks upright like a man and rarely, if ever, moves on all fours. It is generally said to have dark brown, grey, or black fur, but honey-coloured or reddish hair has been reported. Sometimes a long mane of hair that falls down to the shoulders is also mentioned.

The Orang-pendek generally seems to be a solitary creature though there are rare reports of groups of them being seen together.

Many Indonesians fear the Orang-pendek on account of its massive strength, but it is not thought normally of as aggressive. Mostly the creature will move away from any human it sees. It is said to occasionally hurl rocks and sticks as crude weapons when it feels threatened. Like most wild animals, it is probable that the Orang-pendek might become aggressive if cornered or surprised.

Its diet is mainly herbivorous, consisting of fruits, vegetables, and tubers. There are some reports of the animal ripping open logs to get at insect larvae. Rare reports tell of it taking fish and freshwater molluscs. Early reports tell of some of them feeding off the flesh of dead rhinoceros that had fallen into native pit traps.

Native knowledge of the creature goes back into the mists of history and there are a number of localized names for the Orang-pendek around the Island. It is called sedapa or sedapak in the southeastern lowlands. *Gugu* is the name in southern Sumatra whilst in the Rawas district it is *atu rimbu*. In Bengkulu it is known as *sebaba*.

One of the earliest accounts comes from an English employee of the East India Company, William Marsden, accounted in his 1784 book, *The History of Sumatra: Containing an Account of the Government, Laws, Customs and Manners of the Native Inhabitants, with a Description of the Natural Productions, and a Relation of the Ancient Political State of That Island*. In it he mentions a race of rarely seen wildmen covered in long hair.

Indonesia was once a Dutch colony and there are many reports from westerners in the early part of the twentieth century.

In 1918, the Sumatran Governor, L.C. Westenenk, wrote about the creature. He recorded an event which took place in 1910.

> *A boy from Padang employed as an overseer by Mr. van H— had to stake the boundaries of a piece of land for which a long lease had been applied. One day he took several coolies into the virgin forest on the Barissan Mountains near Loeboek Salasik. Suddenly he saw, some fifteen metres away, a large creature, low on its feet, which ran like a man...it was very hairy and was not an orangutan; but its face was not like an ordinary man's...*

Westenenk also recorded another encounter. In 1917, a Mr. Oostingh, owner of a coffee plantation at Dataran, was in the forests at the base of Boekit Kaba when he saw a figure sitting on the ground about thirty feet away.

> *His body was as large as a medium-sized native and he had thick square shoulders, not sloping at all. The colour was not brown, but looked like black earth, a sort of dusty black, more grey than black.*
>
> *He clearly noticed my presence. He did not so much as turn his head, but stood up on his feet; he seemed quite as tall as I (about 1.75m).*
>
> *Then I saw that it was not a man, and I started back, for I was not armed. The creature took several paces, without the least haste, and then, with his ludicrously long arm, grasped a sapling, which threatened to break under his weight, and quietly sprang into a tree, swinging in great leaps alternately to right and to left.*

My chief impression was and still is: 'What an enormously large beast!' It was not an orangutan; I had seen one of these large apes before at the Artis (the Amsterdam Zoo).

It was more like a monstrously large siamang, but a siamang has long hair, and there was no doubt it had short hair. I did not see the face, for, indeed, it never once looked at me.

A Dr Edward Jacobson said that in 1916, while he was camped near the base of Boekit Kaba Mountain, some scouts told him they had seen an Orang-pendek breaking open a fallen log as it looked for insect larvae. When the creature saw the scouts, it ran away on its hind legs. Jacobson also reported that he had seen some footprints at Mount Kerinci. They were like those of a human, but shorter and broader.

Another Dutchman, a surveyor called R. Maier of Benkoelen, had a large collection of footprints. The footprints in Maier's collection had come from Roepit, Boekit Kaba, and Marga Ambatjung. The tracks were made in the late 1910s and early 1920s.

Another Dutch settler, a Mr. van Herwaarden, began researching the creature in 1916 and was initially a sceptic. But in 1918 he found a series of footprints near Moesi Oeloe. Later, he talked to a man called Breikers who had found similar tracks. Van Herwaarden eventually met three Kubu natives who said they had seen an Orang-pendek; it was about four and a half feet tall, they said, with a hairy body, long hair on its head, and long canine teeth.

Some years later, he heard that two corpses were found in the forests near Pangkalan Belai. The bodies were of a female and a child. The Malay who found the two tried to bring the bodies back to civilization, but he was soon forced to abandon them.

Herwaarden saw the animal for himself on the island of Pulau Rimau, cut off by rivers, in October 1923. He had been hunting wild pigs but having failed to find any sat quietly in hiding. After a while something in a tree caught his eye. He described it thus...

It must be a sedapa, I thought. Hunters will understand the excitement that possessed me. At first, I merely watched and examined the beast which still clung motionless to the tree. While I kept my gun ready to fire, I tried to attract the sedapa's attention, by calling to it. But it would not budge. What was I to do? I could not get help to capture the beast. And as time was running short, I was obliged to tackle it myself. I tried kicking the trunk of the tree, without the least result. I laid my gun on the ground and tried to get nearer the animal. I had hardly climbed three or four feet into the tree, when the body above me began to move. The creature lifted itself a little from the branch and leaned over the side so that I could then see its hair, its forehead, and a pair of eyes which stared at me. Its movements had at first been slow and cautious, but as soon as the sedapa saw me, the whole situation changed. It became nervous and trembled all over its body. In order to see it better, I slid down onto the ground again.

The sedapa was also hairy on the front of its body; the colour there was a little lighter than on the back. The very dark hair on its head fell to just below the shoulder blades, or even almost to the waist. It was fairly thick and shaggy. The lower part of its face seemed to end in more of a point than a man's; this brown face was almost hairless, whilst its forehead seemed to be high rather than low. Its eyebrows were the same colour as its hair, and were very bushy. The eyes were frankly moving; they were of the darkest colour, very lively, and like human eyes. The nose was broad with fairly large nostrils, but in no way clumsy. It reminded me a little of a Kaffir's. Its lips were quite ordinary, but the width of its mouth was strikingly wide, when open. Its canines showed clearly from time to time as its mouth twitched nervously. They seemed fairly large to me, at all events they [the canines] were more developed than those of a man. The incisors were regular. The colour of the teeth was yellowish-white. Its chin was somewhat receding. For a moment, during a quick movement, I was able to see its right ear which was exactly like a little human ear. Its hands were slightly hairy on the back. Had it been standing, its arms would have reached to a little above its knees; they were therefore long, but its legs seemed to me rather short. I did not see its feet, but I did see some toes which were shaped in a very normal manner. This specimen was of the female sex and about five feet high.

There was nothing repulsive or ugly about its face, nor was it at all ape-like, although the quick, nervous movements of its eyes and mouth were very like those of a monkey in distress. I began to walk in a calm and friendly way to the sedapa, as if I were soothing a frightened dog, or horse; but it did not make much difference. When I raised my gun to the little female, I heard a plaintive 'Hu-Hu' which was at once answered by similar echoes in the forest nearby.

I laid down my gun and climbed into the tree again. I had almost reached the foot of the bough, when the sedapa ran very fast out along the branch, which bent heavily, hung on to the end, and dropped a good ten feet to the ground. I slid hastily back to the ground, but before I could reach my gun again, the beast was almost thirty yards away. It went on running and gave a sort of whistle. Many people may think me childish if I say that when I saw its flying hair in the sights, I did not pull the trigger. I suddenly felt that I was going to commit murder! I lifted my gun to my shoulder again, but once more my courage failed me. As far as I could see, its feet were broad and short.

In May of 1927, A. H. W. Cramer, a Dutch plantation employee who lived in Kerinci, reported seeing an Orang-pendek from a distance of only thirty-three feet. He was with a native woodcutter at the time. It had long hair and black skin. The beast ran away at high speed leaving small, humanlike foot prints.

In the same year, a sergeant major with the Topographical Service named Van Esch was working in Surulagun. He climbed down a cliff to get some water. He saw a human-like figure approaching him through the jungle and so he hid. As he watched, a four foot two-inch-tall creature came within sixteen feet of his hiding place. It was covered with brown hair and had a chest half a meter wide. It had a long head with a dark face and prominent canine teeth. It was startled by a noise in the forest and ran away.

Also, in 1927, an Orang-pendek was caught in a tiger trap but broke free. The traces of blood it left were examined by zoologist KW Damerman who determined that it was not from a bear, gibbon, or human.

In the 1930s, the interest in the Orang-pendek waned and it was not until the late 1980s and 1990s that the world once more sat up and took notice of Sumatra's short man.

In the 1980s, explorer Benedict Allen first learned of the mystery ape from American businessman called Swartz, who was on a six-country development project tour. Swartz told him that there were stories doing the rounds amongst workers on the Trans-Sumatran Highway of groups of 'monkey-men' hurling sticks at the machinery.

Englishwoman Debbie Martyr began her research in the late 1980s. Debbie first visited Sumatra in July of 1989 as a travel writer. Whilst camped on the slopes of Mount Kerinci at some 11,150 feet her guide Jamruddin pointed out areas were Sumatran rhinoceros and tigers could be seen. Then, casually he commented that in the forested mountains east of Gunung Tujuh Orang-pendeks were sometimes seen. When Debbie made a sceptical comment, Jamruddin told her he had seen the Orang-pendek twice himself. He told her it was still common but becoming rarer in the Kernici area due to the incursions of farmers.

Intrigued, Debbie hit the trail. She found that in the more populated areas around the foothills of Mount Kerinci no one had seen an Orang-pendek in three years. However, people who gathered rattan in the forest had seen them more recently. In lower lying areas, the creature was little known but in remote villages in the hills there were recent reports.

Debbie gathered detailed descriptions from the leaders of villages sometimes as much as sixty miles apart. The Orang-pendek was three to five feet tall with a pot belly, more prominent ears than a siamang gibbon, and a high forehead. It had a mane of hair that could be black, dark yellow, or tan, that hung down to the base of the spine. The body was covered with black or grey hair that was thicker on the limbs. All witnesses said the creature was bipedal.

One witness she spoke to, an old *dukun* or witch doctor living near Mura Amat, said that the Orang-pendek ran with its arms held out. Many reliable sightings came from semi-cultivated areas on the edge of the forest where the animal was seen eating sugar cane and bananas. There were five such reports, one from Mount Kerici near the village of Palompet and the others from settlements around the small town of Lempur, some thirty-three miles southeast of the large town of Sungeipenuh. The sightings occurred before 7 a.m. and after 3 p.m.

One witness, a thirty-three-year-old man from close to Palompet described his sighting thus...

> *I was in my grandfather's house (a bamboo hut) in his fields and looked out and saw two Orang-pendeks. One was bigger than the other. They were eating sugar cane. I went out to look at the more closely. The big one saw me. Then they both ran away. They ran like men. Quite fast.*

The witness stated they were not humans or monkeys. He was offended when Debbie suggested they might be sun bears or siamangs.

All witnesses she spoke to rejected these explanations. They also rejected the idea that what they had seen were the forest-dwelling Kubu people. Orang-utans are native only to the north of Sumatra and are quite distinct from the Orang-pendek. None of the natives attached any mystical significance to the Orang-pendek, they looked on it as a flesh and blood creature.

From there, Debbie travelled to Curup, Bukit Kaba then north to Pedang and Bukitinggi. There was little rainforest in these areas and no one knew of the Orang-pendek. However, upon on her return to England, Debbie learned that there had been sightings around all of these places in former times.

Sufficiently convinced she was onto something, Debbie returned to Sumatra in September and spoke to the head man of the village of Selempaing that lies some 3,600 feet above sea level and around thirty miles south west of Sungeipenuh. The headman said he had twice seen a female Orang-pendek in the fields around the village in recent weeks.

The head man along with a reformed rhino poacher called Musih acted as guides in the treacherous terrain outside of the village. It had no paths and consisted of steep ridges and fast flowing rivers. Only two years before, a tiger had killed a man in the same area. Musih warned Debbie that it might be weeks before they encountered an Orang-pendek but she told him that she would be happy just finding evidence. The men were incredulous when they heard that the orang-sarjana (scientist men) did not know about the Orang-pendek. Musih thought it would be possible to capture an Orang-pendek using a snare or net but said no local people would help if they thought the creature was going to be harmed. He himself said that if one were caught it should be returned to the wild. There seemed to be some kind of taboo about hunting the Orang-pendek.

After a day's trek south, the trio came to a forested river valley at an altitude of 4,600 feet. Here Debbie found the tracks of two Orang-pendek. She followed them up a steep muddy bank for 120 feet before they became lost in a drier area. The larger of the two sets was six inches long and just under four inches wide at the broadest part. They resembled human tracks but were broader and with a larger ball. The instep was heavily imprinted. She calculated that the creature that made them was a little over four feet tall.

She took photos that were of little use due to poor lighting but also took two plaster casts. One was destroyed on the gruelling trek back to Sungeipenuh. The surviving cast she showed to the park director at the headquarters of Kerinci Seblat National Park. Prior to seeing the cast, he had dismissed the Orang-pendek as a folk-tale because the local people were 'simple'. Within an hour of seeing the cast, the director and his deputies admitted that it was not from any animal they knew of.

The cast was sent on to the Indonesian National Parks Department in Bogor. Since then, despite a number of requests, they have not published their conclusions.

Debbie, together with photographer Jeremy Holden, engaged in a fifteen-year search funded by Fauna and Flora International. Jeremy used camera traps set up in remote jungles but failed to capture the creature on film. Jeremy Holden caught a glimpse of the creature as he climbed over a ridge in the jungle. The Orang-pendek swiftly moved away from him. He only saw it from the back but noted it walked upright like a man. Jeremy later moved on but Debbie stayed in Sumatra, becoming the head of the Indonesian Tiger Conservation Group. Since then she has had her own encounters with the short man, these will be covered in the second volume of this work.

My good friend Adam Davies, together with Andrew Sanderson and Keith Townley, found and cast Orang-pendek foot prints and collected hair in the Kerinci area.

Primate biologist Dr David Chivers, of the University of Cambridge, compared the cast with those from other known primates and local animals and said...

> ...the cast of the footprint taken was definitely an ape with a unique blend of features from gibbon, orangutan, chimpanzee, and human. Upon further examination, the print did not match any known primate species and I can conclude that this points towards there being a large unknown primate in the forests of Sumatra.

On March 22, 2010, Tim de Frel and Martijn van Opijnen, both of Den Haag, Holland, interviewed a Sumatran man named Pak Man who claimed to have seen the creature. Pak Man, fifty-four, a resident of Pelompek in Jamba Province said he had seen the Orang-pendek on more than one occasion.

The first part of the interview was done in the morning at the house of Sahar Dimus, a man who would later be my guide on several expeditions in search of Orang-pendek. Sahar functioned as the interpreter.

Pak Man gave an account of a sighting in his shed on his farm, north of Pelompek along the highway. The sighting occurred some ten days previous at around ten o'clock at night.

He was sleeping in his bunk when he heard something breaking into his shed. At first, he thought it was a bear or a tapir, but when he looked through the planks of his bunk, he saw a figure sitting next to a pole, eating from some sugarcane that was in a plastic bag. He could see the animal very well, because a burning candle was at the pole just above the animal. It was about three feet tall.

The animal was covered with short hair, a bit longer in the neck. The features that struck him the most were the large, round torso, a very large stomach, short legs, and long arms. The animal had a rounded head, no neck, bright blue eyes, long eyebrows, two holes instead of a nose, a big non-protruding mouth with thin purple lips and its ears sticking out. No genitals or breasts were visible.

The animal had long fingers, with its thumbs at the side of the hand. It used its hands in the same fashion that humans do. It ate the pieces of sugarcane as humans eat corn. No tail was visible. On each foot it had four average-length toes and one very large one sticking out at the side. The general impression the animal gave was that it had great strength. The respondent was very afraid and stayed quiet for about ten minutes. He immediately knew that it was an Orang-pendek because of the many stories he had heard in the past, some told by his grandfather. He was certain it was not an orangutan, because he had seen one in the zoo of Bukittingi. It was also not a monkey. He later heard that four other people saw a similar animal in the same week.

After ten minutes, he decided he had seen enough and kicked at one of the planks of his bunk. The animal then walked to the side of the shed and tore away a piece of the wall and walked away. The Pak Man said the animal walked bent over a bit, its feet sticking to the sides and its arms hanging in front of its body. It walked with short steps and slowly. Pak Man then grabbed a spear and followed the animal outside. There it vanished in the dark.

Pak Man was then shown some pictures in a book about animals. When showed a picture of an orangutan, he was certain this was not his animal. A picture of a gorilla and a chimpanzee were much more similar to what he had seen.

The second part of the interview was done on the same day, in the shed of the witness. The shed is a forty-five-minute uphill walk from the highway north of Pelompek. It took thirty minutes by motorbike to reach the location. The area is scattered with small plots and little sheds. The plot of the sighting was located a mile from the edge of the forest. The shed consisted of two spaces, one with a mud floor and a small bunk elevated from the floor. The only way in is by a door at the side. There are a couple of small windows without glass.

Pak Man was asked to tell the story again, and some details were added. For instance, the animal had light, almost white hair around its eyes. The pupils were bright blue. The hair on its body had a light violet grey colour. When he scared the animal, it made a sound like a loud cough. Otherwise it was silent. The witness thought the animal was too heavy to climb trees.

The coughing noise is of particular interest as gorilla's employ such a sound as a warning or threat.

The witness and Sahar were certain that the animal was after sugar, because most of the fruit trees in the forest at that time were not bearing fruit. Just outside the shed, where the earth was loose, a footprint was faintly visible. It looked like the print of a human hand and was of the same size. Because it had rained in the previous days, the outline was very vague. The big toe which Pak Man had mentioned, didn't stand out in the print. It looked more like a thumb.

When going back to the road, some large pieces of sugarcane were found, which were broken off from the stem. It looked like they were ripped open by teeth and then sucked at. According to the witness that was the work of Orang-pendek. As the pieces may had been there for days in the rain, and no trash receptacles were available, it was decided to leave the pieces where they were found.

The Orang-pendek is almost certainly a new species of ape. The tracks, with their offset big toe, are indicative of an ape rather than a hominin. Sunda was a large landmass that once incorporated Sumatra, Borneo, Jave, the surrounding islands and the Malayan peninsular and connected them all to mainland Asia. As melting glaciers flooded the oceans nineteen thousand years ago, sea levels rose and the huge landmass became cut into the islands we know today. We know orangutans had already speciated some four hundred thousand years ago. We do not know why this occurred, but the more gracile Sumatran form and the robust Bornean separated. The robust populated the eastern island of Borneo and the gracile the western island of Sumatra. Recently a second species of orangutan has been discovered on Sumatra. The Tapanuli orangutan was recognised as genetically different from the Sumatran orangutan and found farther south.

All this argues for the Orang-pendek as a fourth extant species of orangutan. Newly adapted for a terrestrial rather than arboreal lifestyle.

I will describe my own search for the Orang-pendek in the second volume of this book.

Africa

The original monster ape of Africa can now be seen in any decent sized zoo. You will recall from chapter one that the gorilla was once thought to be nothing more than a hairy ogre from native folklore. Now we know better that there are other, more humanlike creatures reported from the Dark Continent.

The best known of the smaller African humanlike beasts is the *agogwe*. The first recorded sighting of an agogwe by a non-native was documented in 1937 by Captain William Hichens in the December edition of the London magazine *Discovery*. Describing his 1900 encounter with the agogwe in Tanzania, Captain Hichens wrote...

> *Some years ago, I was sent on an official lion hunt in this area while waiting in a forest glade for a man eater, I saw two small, brown, furry creatures come from the dense forest on one side of the glade and then disappear into the thicket on the other side. They were like little men, about four feet high, walking upright, but clad in russet hair. The native hunter with me gazed in mingled fear and amazement. They were, he said, agogwe, the little furry men whom one does not see once in a lifetime.*

Hichens' efforts to follow the hominoids were in vain. According to what villagers later told Captain Hichens, the creatures would weed and hoe people's gardens at night in exchange for food and millet beer. He was ridiculed later for his story, but another westerner had seen what appeared to be the same type of creature over the border in Mozambique.

In support of Captain Hichens story, British officer Cuthbert Burgoyne wrote a letter to *Discovery* magazine in 1938 recounting his personal sighting of two agowes in 1927 while travelling Portuguese East Africa aboard a Japanese cargo boat. Burgoyne wrote...

> We were sufficiently near to land to see objects clearly with a glass of twelve magnifications. There was a sloping beach with light bush above upon which several dozen baboons were hunting for and picking up shell fish of crabs, to judge by their movements. Two pure white baboons were amongst them. These are very rare, but I had heard of them previously. As we watched, two little brown men walked together out of the bush and down amongst the baboons. They were certainly not any known monkey and they must have been akin or they would have disturbed the baboons. They were too far away to see in detail, but these small humanlike animals were probably between four and five feet tall, quite upright and graceful in figure. At the time I was thrilled as they were quite evidently no beast of which I had heard or read. Later, a friend and big game hunter told me he was in Portuguese East Africa with his wife and three other hunters, and saw mother, father, and child of apparently a similar animal species, walk across the farther side of a bush clearing. The natives loudly forbade him to shoot.

Professor A. Ledoux of the Faculty of Science of Toulouse University was the head of the Zoological Department of the Institute of Education and Research at Adiopodoum which was then being formed twelve miles from Abidjan in what was then the Ivory Coast.

One evening in 1947, a young native who worked in his laboratory asked him about African pygmies. The professor told him that they were found in Central and Equatorial Africa and lent him a book on the subject. Ledoux asked him why he wanted to know. He replied, one of his colleagues in another scientific department of the Institute had seen one not far away on the previous day, only five hundred yards away.

When the professor asked why the man had not told him, the native said his friend was afraid he would be laughed at or thought to be mad. The professor promised that he would not make fun of him and would not tell anyone his story.

The next day, he had a visit from the boy responsible for the observation. He was well-educated and had a certificate for primary studies. The professor asked the witness about the pygmy.

> *It happened near the Meteorological set when they were taking their daily readings at eight o'clock in the morning. Amongst the roots of a silk-cotton tree (Bombax) there suddenly appeared a little man with long reddish fur and long hair on his head, the same as a white man but also reddish.*

> *At once the little red man and the large black one took to their heels in opposite directions. For, according to the legends, the little forest men brought bad luck. You only saw them once in a lifetime and you had to be alone.*

> *I went to the place with my two informants. It lay in the shadow of thick forest but was not too overgrown since the silk-cotton tree grew near a path. It was very likely that if there had been anything there it would have been easy to see.*

> *I asked to be informed at once if a similar meeting occurred again, but this never happened.*

It should be noted here that true pygmies have short hair on their heads and very little hair on their bodies at all.

The professor showed him a book containing pictures of Central African pygmies, but the witness insisted that his creature was not like them.

Professor Ledoux later questioned several Africans who trusted him.

> *It's a matter of fact I did not obtain any important information, for while there were plenty of men who 'had seen' them, they were reticent on the subject, always concluding that they were probably mistaken for all the encounters had taken place at nightfall. This is likely enough.*

> *There was one relatively exact fact. In March 1946, a team of workmen under one Djaco who later became my lab-boy and my informant and who died of poisoning in 1949, together with*

a European of whom I can find no trace, were supposed to have seen one of these little red men, at about eight in the morning, in a tall tree in a very wooded little valley about half a mile from the future site of the station. The European asked what it was and the natives explained what a rare thing it was to see such a creature and the evil effects of doing so.

I was at once deluged with stories of dwarfs with their feet back to front, people who lived half in the lagoon and half on land (I think that manatees must be responsible for this legend). These tales were of no interest to me, but I mention them so that the record should be complete.

He then questioned the Europeans who had travelled in the Ivory Coast: One of them told him the following:

During one of his expeditions in the course of 1947, the great elephant-hunter Dunckel killed a peculiar primate unknown to him; it was small with reddish-brown hair and was shot in the great forest between Guiglo and Toulepeu, that is, between the Sassandra and Cavally rivers. Its remains disappeared while it was being carried home, no doubt having been disposed of by superstitious porters. Dunckel even offered to take my informant to the place and he in turn invited me to go with them.

In 1951, the professor's new boy told him that when he was young, probably around 1941, he had himself seen a hunter at Segu bring back a little man with red hair in a cage. The local official had put clothes on it and sent it to Abidjan by way of Bouak. The boy did not know what happened to the little prisoner afterwards.

Professor Ledoux continued...

According to an African technician of mine from Toulepeu called Mehaud Taou, an intelligent boy keenly interested in these questions, there was recently a system of barter between the natives and these forest creatures; various manufactured goods were left in the forest in exchange for various fruits. This was supposed to have gone on until 1935. The little men who practised this barter were hardly known even to the natives themselves. The Gueres called them Sehites.

He concluded that the stories of these pygmies were based on some kind of unknown, red-haired primate.

Could these sightings, in the continent that was the cradle of man, represent a surviving strain of australopithecine? These were a genus of African hominids consisting of eleven species that ranged from 5.6 to 1.5 million years ago. One of them is thought to have evolved into the genus Homo that would eventually give rise to man but there is no consensus as to which species this was.

Could it be possible that one or more species have living dependents that retained the primordial form and have moved away from the grasslands to the jungles of Africa to this day?

North America

Apart from the Yeti, North America's Sasquatch or Bigfoot must be the most famous of all mysterious primates. It is by far and away the most frequently reported. Generally described as seven to ten feet tall, upright, muscular, and with an ape-like face, Sasquatch seems very like the larger kind of Yeti. It's hair ranges from black to brown or reddish. A few white specimens are noted, unlike the Yeti where white-furred individuals are almost unknown.

The creature is reported from nearly every mainland state of the US and from all provinces of Canada. First nation cultures have dozens of names for it including *bukwus, omah, skookum,* and *choanito* to name but a few. The term Sasquatch is derived from a mispronunciation of the Salish *sásq'ets* meaing 'wild man'. J.W. Burns first used the term in the 1920s. Burns spent many years as a teacher on the Chehalis Indian Reserve beside the Harrison River about sixty miles east of Vancouver, BC. Burns collected Indian stories of hairy giants.

The other popular name, Bigfoot, was created by journalist Andrew Genzoli, in a *Humbolt Times* article from October of 1958 titled 'Giant

Footprints Puzzle Residents along Trinity River'. It told of the discovery of huge manlike footprints eighteen inches long discovered by Jerry Crew and his construction team at Bluff Creek, California. The tracks seemed to be made by something that visited the construction site at night and threw heavy objects about.

In 2002, team member Ray Wallace is said to have made a deathbed confession to his family that he had the tracks with carved wooden feet. However, the tracks found and cast at Bluff Creek showed flexibility that would be impossible to recreate with Wallace's crudely made, stiff wooden feet.

In most cases, the creatures seem to avoid humans, perhaps realising how dangerous they are. However, there are a few cases where they have behaved in an aggressive manner, especially if provoked.

In 1924, Fred Beck and four friends were prospecting in a gorge near Mount St Helens, Washington. They found huge, humanlike footprints. Later they caught glimpses of hairy, apish beasts over seven feet tall. Beck shot one that pitched into the gorge. As darkness fell and the men sat in their log cabin, a massive thud hit the side of the building. More followed and the men heard rustling and foot falls from outside. Peering between a gap in the logs they saw three of the creatures outside. They were hurling rocks at the cabin, shaking it violently. The creatures even leapt upon the roof. Luckily the cabin had no windows and was lit by a fire from within. The men shot at the beast through gaps in the logs.

The siege continued until dawn when the creatures left. The men fled, leaving most of their equipment behind.

It has been suggested that some people who have vanished or been killed in wilderness areas were in fact the victims of rogue Sasquatch. One of the stories that makes the hairs on the back of my neck stand on end, took place in Yosemite National Park, California, an area where there have been many odd disappearances. Evelyn Consuela Roseman, a masseuse and stripper from San Francisco, had taken a holiday, hiking alone in Yosemite. On October 19, 1968, three hikers found her battered body near the base of Nevada Fall. Investigators determined that she had been thrown off the five hundred- and ninety-four-foot waterfall. The waters were very low, and she had not been swept off. Her trousers had been pulled down and her top pulled up. The body was so far from the Fall that the investigators concluded she had been thrown rather than fallen. The body had then been dragged some distance and the clothing disturbed. A post mortem reveals she had died from massive internal and cranial damage due to the impact. Her head had been destroyed by the impact as well. More disturbingly, there were bloodless lacerations inside her vagina, showing that the body had been sexually interfered with after death. Semen was found inside the body. The horrific story begs the question just what has the strength to pick up a full grow woman and hurl her so far off the edge. DNA profiling did not exist until the 1980s, but if it had existed back in 1968, I would guess that the samples they took would not be from anything you and I would call 'human'.

I happen to have a good friend who is a Sasquatch witness. Jackie Tonks is a social worker who lives in Bradninch, Devon. She was not a believer in Bigfoot until the following events occurred along a logging track called the Gasquet-Orleans road.

> There were three witnesses. Myself and Thom Cantrall (an author and Big Footer for at least forty years) were in the front of the car and Thom was driving. Aria Cailleach Collett (Bigfoot researcher, author, and musician) was in the back and only caught the end of the activity as I think she was snoozing. Suddenly, from a distance, I saw what looked like two upright figures running in a strange way. They must have been about one hundred feet away and looked rather tall. Initial impressions were that they were men, who appeared to be wearing an all-in-one dark grey overall with the hood up. They had bare feet and I could also see bare faces and hands. I initially got concerned as I knew there was illegal cannabis growing going on amongst the trees and we had been warned by the local ranger to be careful about walking off the paths due to this. I therefore rationalised that the two figures must be drug runners. Their shoulders looked very broad and I initially rationalised they had drugs stashed in a rucksack under the overalls to account for this (which would have been pretty

strange in itself, but you try to make rationale sense of what you see, in terms of your current experience). I said to Thom, something like, 'What the hell are those guys doing, it must be killing their feet running on that barefoot!' They were running in a skittish way, like they were spooked, and I did find this reaction pretty strange. It was like how a deer runs across a road, not a human. The way they were running also looked pretty unusual, they were not moving in the usual way.

At that point Thom laughed and said something like, 'What are you here to look for!' and 'You've just seen them!' However, I then heard him yell and he suddenly started braking dramatically. There they were again, in the middle of the road, but this time a lot closer to us, about forty feet from the car! For about a second and a half they dithered in the middle of the road, not knowing where to go. I presume they had panicked when they could not get tree cover, due to the trees all being burnt out over the other side of the road. For a second, I thought we were going to hit them and shot my arm out defensively in front of me.

I could see what looked like an unfinished fibre glass-like texture on what I had initially thought to be a suit which I later realised was hair (not fur). It was a dark slate grey, but I think they were actually black with dust on them as it was very dry and dusty. One was a bit taller than the other. I would say between about six and a half and seven feet tall (Thom gave an estimate of about thirty inches extra for each of them on top of that and I think he may be likely a better judge as he is ex-forces and also an ex-logger). The top of the head was domed like a gorilla. The face was like a cross between a person and a gorilla. The nose was a proper hooded nose like a human but very flat to the face and quite wide. There were big brow ridges and dark eyes (I could see no white) and they were a little larger than humans and deep-set. The mouth and the jaw were long. The skin was dark olive. The legs were quite long and much thinner in proportion to the rest of the body, and the chest and shoulders very wide, the shoulders about three feet plus.

They were not 'men in monkey suits' as I could see facial muscles clearly twitching in agitation. They made no sound and I could not smell anything. After the hesitation they ran towards the trees again and jumped and ran down a steep slope out of sight. We were so taken aback we just carried on driving (if you read a lot about road sightings of Bigfoot you will note that most people do this; it is as if things are still sinking in and they are still trying to process what they have seen). The next day, we went back but the slope was so steep (almost sheer) that you could not go down to look for footprints. There is no way a human could have run down that without breaking their neck. Sadly, the incident was not filmed as that was the day my camcorder battery was still charging (I had received the wrong duplicate a day before I left for the flight, so did not have a spare)! However, even if I had had it rolling already, it is unlikely I would have caught the incident as it all happened so fast. I now always have an action camera filming constantly whenever in Bigfoot country, but I am sure I won't get such an opportunity again!

For days I tried to rationalise they were just really weird men, but another part of my brain knew what I had seen, but could not come to terms with it. It is like seeing a unicorn in the middle of the road. On about the third day I suddenly went into shock and started shaking and was muttering to myself, 'I've seen Bigfoot!' It is really amazing when you discover they are real! This experience has turned Bigfooting or 'Squatching' into a serious hobby. I try to go out to the Pacific Northwest every year now if I can.

In November of 2017, I also struck luck again. Myself and my friend, Brooklyn Francisco, had attended the Sasquatch Summit, Ocean Shores, Washington. On the Saturday night, we had spoken to Johnny who runs the Summit and he had told us there had been *solid activity a few miles from the Summit from the Casino where the conference was based, and he gave us the rough direction. We drove over and picked a random path and parked nearby and walked down. The time was about 11.15 p.m. by the time we stopped on the path. I am not too sure of the distance, but I could in the remote distance see the heat of the engine still red on the thermal (but I do have an extended range one). Anyway, it was slightly raining and we had our waterproofs on and it was quite cold. I had my hat on with the ear flaps so sadly could not hear too much. Brooklyn and I started to make the usual attractant sounds; wood knocks, whoops, and howls. We did this for about fifteen minutes. I must say, we didn't expect too much to happen.*

Suddenly, Brooklyn said he could hear a slight rustling, and I lifted up my ear flaps, but the level was nothing like what you would expect from a eight hundred pound plus beast! I thought it might be a racoon or a small yearling deer and so I wasn't initially that freaked out. Yearling deers get abandoned by their mothers and will follow you around like puppies and you can even stroke them...they just want a friend! However, we then realised it wasn't a deer when Brooklyn said he heard a deep growl come from well above his head (he is over six feet tall and the ground was relatively flat). There was then a sense of panic and we started swinging our torches wildly around to try to get a look at what it might be as it was only a few feet away behind dense undergrowth. My initial fears were that it might be a bear. Suddenly a pine cone landed at our feet. I quickly rationalised it had fallen from a tree and said we should be careful not to over-interpret a random tree fall. However, shortly after a small stone was thrown and flew about four inches in front of my face! At that point we freaked. We instantaneously went into that primal defence of back to back and Brooklyn pulled out his machete. He shouted something like, 'Who's that, stop messing with us!'. I wildly flashed the torch around and tried to use my thermal but the undergrowth was so thick I could see nothing.

The machete I think was a big mistake. A rock about the size of small cinder block (in UK we call them a breeze block) came hurtling out from about four and a half feet away to our left, with such force that it went about a third to a half of the way into the ground and smashed every small stone it hit! Some of the mud from the impact hit my leg. We were out of there! Unless a random human was sitting with a catapult in the middle of winter down a random logging road waiting for a random tourist, then there was no way that was a human! It was thrown with tremendous force. I have, however, seen Silverback Mountain Gorillas in Africa have such force when throwing objects. We ran like crazy back to the car. I learned from that incident: never go Sasquatch hunting with loose trousers...mine started to fall down on the run back!

The most famous piece of evidence for the creature was a sequence of film shot by Roger Patterson and Bob Gimlin on October 20 at Bluff Creek, California. The men had been searching for Bigfoot and were on horseback as they turned a corner into the creek. They saw a tall, bipedal, hairy creature that walked calmly away. Patterson began to record it with his film camera. Patterson's horse reared and threw him to the floor. Scrambling to his feet he followed the creature, still filming. The creature looked back once before walking away into the forest.

Whole books have been written on this sequence of film, the best-known film of a mysterious creature ever taken. It has been discussed and analysed by various 'experts' from around the world, often to completely different conclusions. Here I will share with the reader my own thoughts, looking at the film through the eyes of a natural historian.

Firstly, the creature is a female, with visible breasts. If you were going to fake a film of Bigfoot by using a tall man in an ape suit, where in the equation would you think of adding large, hairy breasts? Such an artefact would make the costume more expensive and harder to create. In known ape species, the females have fairly flat breasts. Human females have rounded breasts as a counterbalance to the large buttocks. Humans are bipeds and walk upright. The gluteal muscles keep the body level when the legs are lifted. Human female pelvic girdles are broad in order to accommodate the birth canal. Hence the buttocks of human females are larger and more rounded then males. Great apes are knuckle walkers and move on all fours, ergo they lack developed buttock muscles and the pendulous breasts that counterbalance them. A hypothetical upright-walking female ape would have rounded buttocks and breasts. The creature in the Patterson-Gimlin film possesses both of these.

Secondly, the creature turns its head sideways and the viewer can clearly see a thick brow ridge and a forehead that slopes away in an angle to make a cone-shaped head somewhat like that of a gorilla. Fossil hominins display this same acutely sloping forehead above a thick brow ridge. The human forehead rises up directly above the brow. If the creature in the film was a man in a suit his human head would not fit into a mask with such a sloping forehead, there simply would not be enough room unless the mask was very oversized like some kind of carnival head piece which it is clearly not.

Thirdly, the limb and body proportions of the creature are non-human. The torso is longer than a human's and the hips proportionately lower. The arms are 10 percent longer than a human. The upper legs are longer than a human and the lower part is shorter. Even if you could make such a convincing costume you could not get a human to fit into it. The arm and leg joints cannot be made to line up. Muscles can clearly be seen moving under the hair. In short, the subject in the film is not a man in a costume.

And then we have the footprints. Jimmy Chillcutt, a crime scene investigator and latent fingerprint examiner from the Conroe, Texas Police Department, has taken fingerprints of many primates in zoos. He has examined many of the Sasquatch print casts in the collection of Jeff Meldrum. He has found dermal ridges that lay parallel to the edge of the feet.

> *The ridge flow pattern and the texture were completely different from anything I've ever seen. It certainly wasn't human, and of no known primate that I've examined. The print ridges flowed lengthwise along the foot, unlike human prints, which flow across. The texture of the ridges was about twice the thickness of a human, which indicated that this animal has a real, thick skin.*

The Sasquatch seems very much like a larger kind of Yeti. It could be that they are sub-species. Asia and North America were once connected by a land bridge known as Beringia that crossed what is now the Bering Strait. Beringia allowed animals such as mammoths to cross over from Asia to North America. *Gigantopithecus blacki* may have used this land bridge to expand its range into the New World.

South and Central America

If old world apes could have colonised North America, then there is no reason why they should not have continued south through the continent and colonised Central and South America. Indeed, there are stories of encounters with giant apes and smaller hominins in the jungles of the neo-tropics.

Italian archaeologist Pino Turolla was told a bloody story by an Indian guide in Marirupa Fall, Venezuela in 1968. Antonio, the man in question, had gone with his two sons to the Pacaraima Range. As they approached the savannah, three lumbering, ape-like beasts with smallish heads and long arms attacked them with clubs, killing his younger son. Some six months later, Turolla persuaded Antonio and some other Indians to show him the area in question. When approaching it, they heard shrill roars and the natives would go no further. Turolla himself claimed to have glimpsed an eight-foot, ape-like, lumbering form.

Whilst searching for giant anacondas in Guyana back in 2006, I was told a remarkable tale by Damon Corrie, an Arawak chief. Just three years earlier, two native children, a brother and sister, were walking home from school some miles from the village of Pakuri (they lived in the next village). A huge, hairy man emerged from a stand of trees and seized the girl, who was never seen again. The police did not bother to investigate since the victim was 'only a native'. These Yeti-like beasts are called *di-di* in Guyana.

In 1987, Gary Samuels (a mycologist studying under a grant from the New York Botanical Garden) was examining fungi in Guyana. Hearing footsteps nearby, he glanced up, expecting to see his Guyanese assistant.

Instead, he saw a bipedal, ape-like creature standing about five feet tall. Samuels said the creature bellowed at him, then ran away.

Charles Brown, a British surveyor, collected native reports of the creatures in 1868. In one case, two of the creatures attacked a woodcutter who fought them off with his axe despite being badly cut.

A British magistrate called Haines, a resident in Guyana, saw two di-di in 1910 near Rupununi. He was prospecting for gold along the Konawaruk, a tributary that joins the vast Essequibo River just above its junction with the Potaro. He came upon two creatures that stood up on their hind legs. They had humanlike faces but were covered in a reddish-brown hair. The pair retreated without ever taking their eyes off Haines.

In 1931, an Italian expedition led by Dr Rezo Giglioli, heard more stories of the creatures in the same area. Their native guide, Miegam, said that in 1918, whilst going up the Berbice, he saw what he thought were men on a sandbank in the distance. He called out to them asking how the fishing was, but the figures ran into the jungle. Upon reaching the sand bank he and his friends discovered the footprints there were those of apes, not men.

Claus U. Oheim, an animal dealer from Ecuador, told cryptozoologist Ivan T Sanderson the following story.

> The so-called Shiru I have heard of from the Indians and a few white hunters on both sides of the Andes, but decidedly more so on the eastern slopes, where the vast mountains are still quite unexplored, and rarely, if ever visited. All reports describe the Shiru as a small (four to five feet tall) creature, decidedly hominid, but fully covered with short, dark brown fur. All agreed that the Shiru was very shy, with the exception of one Indian, who claimed having been charged after having missed with his one and only shot from a muzzle loading shotgun, a weapon still used by the majority of Indians along with the blowgun.

The duende is a type of hairy goblin widespread in the folklore of the Neo-tropics. Ivan T Sanderson wrote in 1961 of a sighting in the Central American country of Belize...

Dozens told me of having seen them, and these were mostly men of substance who had worked for responsible organisations like the Forestry Department and who had, in several cases, been schooled or trained either in Europe or the United States. One, a junior forestry officer born locally, described in great detail two of these little creatures that he had suddenly noticed quietly watching him on several occasions at the edge of the forestry reserve near the foot of the Maya Mountains. These little folk were described as being between three foot six and four foot six, well-proportioned but with very heavy shoulders and rather long arms, clothed in thick, tight, close brown hair looking like that of a short-coated dog; having very flat yellowish faces but head hair no longer than the body hair except down the back of the neck and midback.

The creatures are often said to hold palm leaves over their heads like hats, a behaviour seen in chimpanzees. They leave small but deep tracks with narrow heels much like the Orang-pendek.

The *salvaje* are reported from southern Mexico, down through Columbia and into Venezuala. It is here that the legend seems strongest. The salvaje are three to five feet tall and covered in reddish hair. They walk erect and emit cries that sound like a human in distress. The creatures, like many other wildmen, are credited with kidnapping human females.

A Venezuelan hunter, Fernando Nives, claimed to have seen three salvaje whilst tracking game along the Orinoco, twenty-five miles north of Puerto Ayachuco in 1980.

In 1985, a construction worker was bulldozing a road through some forest near Puerto Ayachuco when he saw one of the creatures.

Australia

Apart from bats, rodents, and domestic animals, modern man is the only placental mammal that lives on the continent of Australia. All others are marsupials or monotremes (egg-laying mammals). Strangely, however, there is an ancient legend of huge, gorilla-like beasts haunting the outback.

As with other parts of the world, the aboriginal people knew of them and had many names for the creatures such as *quinkin*, *doolaga*, *jurrawarra* and, most famously, *yowie*.

The Aborigines were said to have fought with the yowies in times past. An elderly man called Harry Williams, a member of the Ngunawal people, said that as a boy, around 1840, he saw a large group of warriors manage to kill one. This occurred on a hillside below the junction of the Yass and Murumbidgee rivers, New South Wales. He said it was black-skinned and covered all over with grey hair.

White settlers also encountered large, ape-like monsters in the bush as far back as the late eighteenth century. Sydney Wheeler Jephcott, a bushman and poet, wrote to the *Sydney Morning Herald* in 1912 about an encounter a friend of his had had at Packer's Swamp, Creewah, New South Wales:

> *I heard that George Summerell, a neighbour of mine, while riding up the track which forms a shortcut from Bombala to Bemboka, had that day, about noon, when approaching a small creek about a mile below Packer's Swamp, ridden close up to a strange animal, which, on all fours was drinking from the creek. As it was covered with grey hair, the first thought that rose to Summerell's mind was: 'What an immense kangaroo'. But, hearing the horses' feet on the track, it rose to its full height of about seven feet and looked quietly at the horseman. Then, stooping down again, it finished its drink, and then, picking up a stick that lay by, it walked steadily away up a slope to the right or eastern side of the road and disappeared amongst the rocks and timber one hundred and fifty feet away. Summerell described the face as being like that of an ape or man, minus forehead and chin, with a great trunk all one size from shoulders to hips, and with arms that nearly reached its ankles.*

Jephcott later investigated and found huge manlike tracks that he cast with plaster from Paris.

Seventy-six-year-old Bob Mitchell of Radcliffe told the *Brisbane Sunday Mail* of his encounter in 1928. He was with two other men in the bushland along the Queensland and New South Wales border.

It was about 10 a.m.—the yowie was standing in a clearing not far from us and in that light, there was no mistaking it for something else. It was about seven feet tall, with a black human face and a gorilla-like body covered in thick brownish hair. It showed no aggression: it just looked at us for a moment, then turned and disappeared back into the bush. It had really big feet and could move fast.

Two weeks later, the same three men saw another *yowie* near the location of the first sighting. It, too, looked at the men for a while before moving off into the bush.

Sightings of yowies have persisted down the years and are not just reported by laymen. In March of 1978, a Queensland National Parks and Wildlife ranger was searching for what he thought was an escaped pig. As he approached, there was a noise in some of the trees, so he looked at the ground for pig tracks but found none. Something made him look up and about twelve feet away stood a hair-covered creature with its hand wrapped around a sapling.

The man, who wished to remain anonymous, spoke to Frank Hampson of the *Gold Coast Bulletin*. He was close enough to see the hair on the back of the creature's manlike fingers. Its skin was black, and it stood eight feet tall. It had a flat, black, shiny face with yellow eyes and a gaping hole for a mouth. Its neck was short and thick. The body was muscular and covered in short black hair. The pair stood frozen, staring at each other for ten minutes before the yowie released a foul smell that made the man vomit. It moved off sideways into the bush.

The witness believed he had seen the creature only because he was moving quietly, as he thought he was stalking a pig.

A politician would risk extreme ridicule or even the loss of his job if he admitted to seeing a giant 'ape-man' in the Australian outback. But that is just what happened to Senator Bill O'Chee of the National Party back in October of 1977, when he was a boy. O'Chee had been part of a party of children from Southport School who had been on a two-day camping trip near Springbrook, Queensland. O'Chee was brave enough to tell the strange story in full. He was interviewed by legendary Australian cryptozoologists Tony Healy and Paul Copper. They also spoke to his friend Craig Jackson.

The camp was called Koonjewarre and stood on open, grazing land close to a dense forest. There were thirty boys aged twelve to thirteen and two teachers. They were all stopping in cabins. At 12.30 p.m., Bill and Craig spotted something lying uphill from the camp, about one thousand three hundred feet away in an open area. After a while it stood up. The boys passed a pair of binoculars between them as they watched it. The creature was huge, about ten feet tall. It was covered in black or dark brown hair about two inches long. It had no neck, the head sitting squarely on the shoulders. Craig said the creature reminded him of the character Chewbacca from the *Star Wars* films except its fur was shorter and it had broader shoulders and was bulkier. The thing swayed from side to side and seemed to be looking around. It has a stooped posture with long arms that fell down past the knees. Close by was a bush covered in white flowers. The two lads noted that it came up to the creature's waist.

Other boys came over and a total of about twenty saw the yowie. Two boys stepped outside and the yowie noticed them and ran for the trees. The children told one of their teachers, Kevin Brooks, an ex-soldier, and he decided to lead a party up the hill to look for the monster. Craig said that when the camp caretaker heard this, he looked really frightened and urged them not to go up the hill. Craig thought that the caretaker had seen the monster before and was scared of it.

Later Healy and Cropper confirmed this by talking to the camp's present caretaker who said that his predecessor had indeed encountered a yowie.

Ignoring the caretaker's warning, Mr Brooks and four brave boys, including Bill and Craig, ventured up the mountain. They found that the bush that had been waist-high on the yowie was five feet tall, making the monster eight to ten feet high. Armed only with sticks, they entered the forest and found a trail of broken saplings and trampled bushes. They also found an area of compressed grass and twigs where the thing had apparently slept.

Back at the camp, the other boys saw the monster emerge from the forest close to where the party had entered and it moved along the edge of the woods. The beast was glimpsed intermittently all afternoon.

That night, the yowie returned, making awful noises and coming as close as thirty-three feet to the cabins. Craig admitted to being 'shit scared'.

In the morning, they found tracks and three-foot-high, deeply rooted shrubs that had been torn out of the ground by something far stronger than a man.

Back at the school, the headmaster asked the boys not to talk about the incident and censored an article about it in the school magazine. Angered by this, Bill contacted the *Gold Coast Bulletin* and told them the whole story.

How could such creatures get to Australia in the first place? The only primates know on the continent are humans themselves. There is a naturally occurring mode of animal distribution known as 'rafting'. Rafting happens when animals are swept out to sea on mats of vegetation and tree trunks during natural disasters such as flooding or tsunamis. These masses of vegetation can be acres in size and contain pools of rainwater and food sources such as fruit or fish. Many animal and plant species have been dispersed in this manner. In the 2004 tsunami that swept across Asia, one man survived for over a week on one of these 'rafts'.

Fossil bones and stone tools belonging to *Homo erectus* have been found on the island of Crete. However, Crete broke off from the mainland long before *Homo erectus* evolved. These hominins did not have the technology to build boats, so they must have arrived on Crete by accident via the rafting dispersal method.

During the ice ages, the sea levels were lower and mainland Asia was much closer to Australia than it is today. The ancestors of the yowie, unknown, upright apes or hominins may have been swept over by a series of natural events to begin a colonisation of the continent by non-human primates long before the first modern humans arrived.

The evidence is now growing for a number of upright-walking apes and relic hominins unknown to science, existing around the world. Traditionally, bipedalism is thought to have developed on the plains of

East Africa, when hominids first left the jungles to exploit new food sources around five million years ago. Standing erect, so the theory said, gave them a better view of potential predators. The vervet monkey (*Chlorocebus pygerythrus*), demonstrates this kind of behaviour, rearing up to look about it for danger. But now it seems that bipedalism may have begun to evolve in the jungles.

During a year-long study of the Sumatran orangutans of Gunung Leuser National Park, palaeoanthropologist Susannah Thorpe of the University of Birmingham spotted apes in the trees a total of 2,811 times, including numerous instances where they walked erect. In 75 percent of these cases, they maintained balance with their hands, and over 90 percent of the time their legs were stiff, unlike the bent-knee, bent-hip shuffle of chimps and gorillas, which also stand upright in trees sometimes.

The apes stood erect mainly to reach for fruit whilst on fairly narrow branches. Thorp postulated that the straight-legged posture helped them balance in the same way as a gymnast on a trampoline. Palaeoanthropologist Bernard Wood of The George Washington University in Washington, DC, commented on the findings.

Most of us had assumed that the only place where it's sensible to be bipedal is on the ground. A handful of fossil species dating from five million to twenty-eight million years old, mostly before chimpanzees split from hominins, showed signs of upright posture and bipedalism, but the evidence has been pretty flaky.

Wood thinks the findings put these fossils in a new light and they may have been true bipeds who evolved to bipedalism to reach for fruit. As the jungles shrank, they took up bipedal walking on the ground whilst the gorillas and chimpanzees took up knuckle walking.

The fossils in question were of course African, but could something similar have occurred in the jungles of Asia, ultimately giving rise to a number of bipedal ape species? Taken with the new fossil and genetic evidence for late surviving hominins and new unsuspected hominid species, the reports of hairy wildmen from around the world look a lot more likely.

CHAPTER SIX

The Magic Zoo

'Why, sometimes I've believed as many as
six impossible things before breakfast.'
—The White Queen, *Alice in Wonderland*,
Lewis Carroll

•

Around the world, in many cultures, there are stories of fabulous animals. These strange beasts come in a wide array of fantastical forms. Could some of them be based on real life creatures, distorted by mythology? Some known creatures have astounding attributes and others have had strange powers attributed to them quite unjustly.

The giant monkey tree frog (*Phyllomedusa sauvagii*) is a large tree frog from South America. Local tribes swore its secretions abated feelings of hunger and thirst, and made hunters invisible in the forest. When biochemists analysed its discharge, they found it did indeed stifle feelings of hunger and thirst, and also masked human odour, in effect, making a hunter invisible.

The hero shrew (*Scutisorex somereni*) of Central Africa was thought to possess magical strength. Natives wore it's pelt on their belts in the belief it could give the wearer its magic might. In fact, the little creature can withstand one hundred and fifty pounds of weight on its back. The vertebrae of the hero shrew are thick, corrugated cylinders that interlock on their sides and lower surfaces. The animal's spine has bony projections that mesh to form a very strong, yet flexible backbone. The result is truly astounding strength.

Gorillas were once thought to carry off and rape native girls and beat elephants to death with tree branches. Of course, they do no such things.

The Kenyan sand boa (*Gongylophis colubrinus*) is a small constricting snake widespread in eastern and northern Africa. The people of Sudan call the sand boa *apris* and are convinced that it is so venomous that even if you touched the snake, it's venom would seep through your skin and kill you. The power, of course, is apocryphal, but the snake exists.

Bernard Heuvelmans called these latter types examples of the Mythification process. Some supposed mythical beasts may have been real creatures that man has given strange attributes to. In a previous chapter, we have looked at the possible links between the Almasty and the trolls of Scandinavian legend. What else may have a basis in fact?

The Unicorn

Apart from the dragon, the unicorn must be the most recognisable of all legendary beasts. Most people think the unicorn was a white horse with a single, spiralled horn. A closer look at medieval pictures of this beast and its many representations in heraldry will show it is something quite different. The unicorn has cloven hooves like a goat or antelope, not singular hooves like a horse. The unicorn was not and never was a true horse. Stories of these creatures reach back centuries.

Ctesias, a Greek physician and historian who lived in the fifth century BC, wrote of the beast saying that it had a horn a cubit long. A Greek cubit was eighteen inches.

Another Greek historian, Herodotus (484 BC–425 BC), writes on the unicorn...

> For the eastern side of Libya, where the wanderers dwell, is low and sandy, as far as the river Triton; but westward of that, the land of the husbandmen is very hilly and abounds with forests and wild beasts, for this is the tract in which the huge serpents are found, and the lions, the elephants, the bears, the aspicks, and the horned asses.

Pliny the Elder (23–79 AD) was a Roman historian, naval commander and naturalist. In 77 AD he published his best-known work, *Natural History*. He includes the unicorn in this book.

> The Orsæan Indians hunt down a very fierce animal called the monoceros, which has the head of the stag, the feet of the elephant, and the tail of the boar, while the rest of the body is like that of

the horse. It makes a deep lowing noise, and has a single black horn, projecting from the middle of its forehead, and two cubits in length. This animal, it is said, cannot be taken alive.

It seems that Plinny is speaking of the Indian rhinoceros (*Rhinoceros unicornis*). Note that it even has unicorn in the second part of its binomial.

Claudius Aelianus (175–235 AD) was a Roman author who wrote widely on animals. He indicated that the country north of the Himalaya, Tibet, and Tartary, still had the reputation of being one of the homes of the unicorn.

Both historians, and the more learned of the Indians, amongst whom the Brahmins may be specified, declare that there is a countless number of these beasts. Amongst them they enumerate the unicorn, which they call cartazonon, and say that it reaches the size of a horse of mature age, possesses a mane and reddish yellow hair, and that it excels in swiftness through the excellence of its feet and of its whole body. Like the elephant, it has inarticulate feet, and it has a boar's tail; one black horn projects between the eyebrows, not awkwardly, but with a certain natural twist, and terminating in a sharp point. It has, of all animals, the harshest and most contentious voice.

It is said to be gentle to other beasts approaching it, but to fight with its fellows. Not only are the males at variance in natural contention amongst themselves, but they also fight with the females, and carry their combats to the length of killing the conquered; for not only are their bodies generally imbued with great strength, but also, they are armed with an invincible horn. It frequents desert regions and wanders alone and solitary. In the breeding season, it is of gentle demeanour towards the female, and they feed together; when this has passed and the female has become gravid, it again becomes fierce and wanders alone.

They say that the young, while still of tender age, are carried to the King of the Prasians for exhibition of their strength, and exposed in combats on festivals; for no one remembers them to have been captured of mature age.

He may well have been referring to the Indian rhinoceros as well.

This explanation cannot be used for the following account. It comes from the writings of John Harris (1666–1719), an English writer, scientist, and Anglican priest. In his book *Complete Collection of Voyages and Travels*, he speaks of the unicorn as inhabiting the mountains of Ethiopia.

The unicorn is found in the mountains of high Ethiopia. It is of an ash colour and resembles a colt of two years old, excepting that it has the head of a goat, and in the middle of its forehead a horn three feet long, which is smooth and white like ivory, and has yellow streaks running along from top to bottom.

This horn is an antidote against poison, and it is reported that other animals delay drinking till it has soaked its horn in the water to purify it. This animal is so nimble that it can neither be killed nor taken. But it casts its horn like a stag, and the hunters find it in the deserts. But the truth of this is called in question by some authors.

German orientalist and scholar of Ethiopian history Hiob Ludolf (1674–1704) mentions unicorns in that country as well.

Here is also another beast, called arucharis, with one horn, fierce and strong, of which unicorn several have been seen feeding in the woods.

In the book *The Navigation and Voyage of Lewes Vertomannus, of Rome, into Arabia, Egypt in 1503*, the author speaks of unicorns he saw kept at the temple in Mecca.

On the other part of the temple are parks or places enclosed, where are seen two unicorns, named of the Greeks monocerotæ, and are there showed to the people for a miracle, and not without good reason, for the seldomness and strange nature. The one of them, which is much higher than the other, yet not much unlike to a colt of thirty months of age; in the forehead groweth only one horn, in manner right forth, of the length of three cubits. The other is much younger, of the age of one year, and like a young colt; the horn of this is of the length of four handfulls.

This beast is of the colour of a horse of weesell colour, and hath the head like a hart, but no long neck, a thynne mane hanging only on the one side. Their leggs are thin and slender like a fawn or hind. The hooves of the fore-feet are divided in two, much like the feet of a goat. The outer part of the hinder feet is very full of hair.

This beast doubtless seemeth wild and fierce, yet tempereth that fierceness with a certain comeliness. These unicorns one gave to the Sultan of Mecha as a most precious and rare gift. They were sent him out of Ethiope by a king of that country who desired by that present to gratify the Sultan of Mecha.

The description here, cloven hooves, mane, and slender legs are not the features of a rhino.

In the following account, we see unicorns listed alongside and quite separately to rhinos. It comes from the accounts of the travels of Johann Grueber, a Jesuit of the seventeenth century.

Sining is a great and populous city, built at the vast wall of China, through the gate of which the merchants from India enter Katay or China. There are stairs to go atop of the wall, and many travel on it from the gate at Sining to the next at Soochew, which is an eighteen days' journey, having a delightful prospect all the way, from the wall, of the innumerable habitations on one side, and the various wild beasts which range the desert on the other side.

Besides wild bulls, here are tigers, lions, elephants, rhinoceroses, and monoceroses, which are a kind of horned asses. Thus, the merchants view the beasts free from danger, especially from that part of the wall which, running southward, approaches Quang-si, Yunnan, and Tibet; for at certain times of the year they betake themselves to the Yellow River, and parts near the wall which abound with thickets, in order to get pasture and seek their prey.

Another Jesuit who claimed to have seen unicorns was Father Jerome Lobo who traveled from Portugal to Abyssinia in the year 1622.

In the province of Agaus has been seen the unicorn; that beast so much talked of and so little known. The prodigious swiftness with which the creature runs from one wood into another has given me no opportunity of examining it particularly; yet I have had so near a sight of it as to be able to give some description of it.

The shape is the same with that of a beautiful horse, exact and nicely proportioned, of a bay colour, with a black tail, which in some provinces is long, in others very short; some have long manes hanging to the ground. They are so timorous that they never feed but surrounded with other beasts that defend them.

Deer and other defenceless animals often herd about the elephant, which, contenting himself with roots and leaves, preserves the beasts that place themselves, as it were, under his protection, from the others that would devour them.

So, what are we to make of the unicorn? It is a one-horned beast resembling an antelope, goat, or deer, with cloven hooves. Some have tried to equate it with the Arabian oryx (*Oryx leucoryx*), a white-coated Middle Eastern antelope. The oryx, however, bears two horns that curve backwards from its head, not a single one pointing forwards.

A better bet for the origin of the unicorn is a prehistoric antelope called *Procamptoceras brivatense*. Procamptoceras lived over much of Europe until around eight hundred thousand years ago. In life, the animal had two horns, one behind the other on its forehead. But these horns were so close together that they would have been covered by a single sheath giving the appearance of one horn. It is possible that Procamptoceras lingered on in relic populations that may have co-existed with man. It is also possible that there were closely related species, with the same morphology in other parts of the world.

The sale of narwhal tusks in the Middle Ages as unicorn horns may have added to the legend. The narwhal (*Monodon monoceros*) is a small

Arctic whale, the males of which bear a tooth that grows into a spiralling tusk reaching five to ten feet long. These were passed off as unicorn horns in the past and changed hands for vast sums of money. The unicorn horn was supposed to be able to negate the effects of any poison.

But there is one last twist in the story of the unicorn, a man-made one! William Franklin Dove (1897–1972) was an American biologist who worked at the University of Maine. Dove conducted research into the growth of horns. He discovered that horns did not grow straight out of the skull; instead, horn tissue developed separately and fused to the skull as it grew. The horn 'buds' were soft and malleable. If drawn together they would form a single horn growing up from the animal's forehead. This only worked with true horns like those on goats and bulls, not on seasonal antlers such as deer possess. Dove's research showed that tribal people in Asia and Africa had practised this and created one-horned goats that swiftly grew to dominance of a herd and could better protect females and young from predators. These animals closely resembled the unicorns of legend. Since this forgotten technique was rediscovered by Dove, a number of latter-day unicorns have been created. A mythical creature resurrected by science!

The Griffon

A legendary beast with the head, talons, and wings of a giant eagle and the hind legs and tail of a lion, the griffon, griffin, gryphon, or grypes, is found in mythology and art reaching back to 3000 BC. In medieval Europe it was associated with Christ, as it combined a creature of the air with a creature of the land as the clerics believed that Christ combined the heavenly with the earthly. It was thought that griffons mated for life and so the church used them as a symbol of their opposition to remarriage.

Herodotus says of griffons in his work *The Histories*:

> But in the north of Europe there is by far the most gold. In this matter, again I cannot say with assurance how the gold is produced, but it is said that one-eyed men called Arimaspians steal it from griffins. But I do not believe this, that there are one-eyed men who have a nature otherwise the same as other men. The most

outlying lands, though, as they enclose and wholly surround all the rest of the world, are likely to have those things which we think the finest and the rarest.

Claudius Aelianus, in his book *On Animals*, records the following.

I have heard that the Indian animal the grupa (gryphon, griffin) is a quadruped like a lion; that it has claws of enormous strength and that they resemble those of a lion. Men commonly report that it is winged and that the feathers along its back are black, and those on its front are red, while the actual wings are neither but are white. And ktesias (ctesias) records that its neck is variegated with feathers of a dark blue; that it has a beak like an eagle's, and a head too, just as artists portray it in pictures and sculpture. Its eyes, he says, are like fire. It builds its lair amongst the mountains, and although it is not possible to capture the full-grown animal, they do take the young ones. And the people of baktria (bactria), who are neighbours of the Indians, say that the grypes (Gryphons) guard the gold in those parts; that they dig it up and build their nests with it, and that the Indians carry off any that fall from them. The Indians, however, deny that they guard the aforesaid gold, for the grypes have no need for it—and if that is what they say, then I at any rate think that they speak the truth—but that they themselves come to collect the gold, while the grypes, fearing for their young ones, fight with the invaders. They engage, too, with other beasts and overcome them without difficulty, but they will not face the lion or the elephant. Accordingly, the natives, dreading the strength of these animals, do not set out in quest of the gold by day, but arrive by night, for at that season they are less likely to be detected.

Now, the region where the grypes live and where the gold is mined is a dreary wilderness. And the seekers after the aforesaid substance arrive, a thousand or two strong, armed and bringing spades and sacks; and watching for a moonless night they begin to dig. Now if they contrive to elude the grypes they reap a double advantage, for they not only escape with their lives but they also take home their freight, and when those who have acquired a special skill in the smelting of gold have refined it, they possess immense wealth to requite them for the dangers described above. And they return home, I am told, after an interval of three or four years.

The Greek poet and traveller Aristeas, who lived some 700 BC, visited Scythia and the area of Eurasian stepp. The Scythians, a race of nomads he met near the Altai Mountains, told him of griffons that guarded gold and forts with horse-riding nomads in the wilderness. They described them as lion-sized, four-footed, and with sharp, strong beaks.

Saint Isidore of Seville (560–635 AD), Archbishop of Seville for thirty years, noted:

The griffin (grypes) is so called because it is an animal with feathers and four feet. This kind of wild animal is born in the Hyperborean mountains. They are lions in their entire torso, but they are like eagles in their wings and faces. They are violently hostile to horses. They also tear humans apart when they see them.

A number of mummified bodies were found in tombs in the Altai Mountains by Soviet archaeologist Sergi Rundenko in the 1940s. Many bore tattoos of griffons. In the 1990s, more mummies with these designs were uncovered in the area.

These Scythian nomads lived along the trade route from China to the West that rand through the western Gobi Desert. Greek and Roman trade flourished here until around the third century AD. The Greeks may have gotten their idea of the griffon from much further east.

In her book *The First Fossil Hunters*, Adrienne Mayor postulates that the griffon legend may have been based on fossils of the dinosaur *Protoceratops andrewsi*. This quadrupedal dinosaur was about as large as a sheep and bore a heavy beak. An herbivore, its fossils have been found across the Gobi in Mongolia in strata rich in gold. Fully preserved skeletons have been found, many in life-like postures. Roy Chapman Andrews, an American naturalist, visited the area and found *Protoceratops* skeletons, as well as exquisitely preserved nests with fossil eggs. The latter are now known to have come from a different dinosaur, *Oviraptor philoceratops*. Andrews wrongly thought that Oviraptor was trying to steal the Protoceratops eggs, but we now know the eggs belong to the Oviraptor who was sitting on the nest at the time of its death and preservation. However, the Protoceratops fossils and the Orviraptor eggs were found in the same area.

Early travellers may have seen the skeletons and eggs of these dinosaurs and thought them to be only recently dead due to the fine state of preservation. The rocks in the area are also rich in gold so we may have the genesis of the four-legged, beaked monster that dug its nest in gold.

Some have also suggested that the bearded vulture (*Gypaetus barbatus*), also known as the Lamergeier or ossifrage, has added to the legend of the griffon. This massive vulture can measure ten feet across the wings and is found in eastern Europe, Asia, and parts of Africa. It has a thick ruff of feathers around its neck that resembles a lion's mane. Up to 90 percent of its diet consists of bones which it can swallow whole or, in the case of large bones, it drops them from a height in order to smash them open and get at the marrow. Unlike most other vultures, it will attack and kill live prey quite savagely. Bearded vultures have been known to kill mammals as large as a chamois (*Rupicapra rupicapra*), a type of mountain goat. They also kill tortoises by dropping them from great heights to crack their shells. The Greek writer of tragedies, Aeschelyus, was killed by a bearded vulture who mistook his bald head for a rock and dropped a tortoise on it. Ironically, his best-known work was *Prometheus Bound* in which the titular character, a titan who stole fire from Zeus, is punished by being chained to a rock whilst an eagle ate his liver.

The Basilisk

The name 'basilisk' means 'little king'. The creature was reckoned to be the king of serpents. It was described as a tiny serpent about a foot in length, bearing a crown or crest upon its head to denote it as the king of serpents. The deserts of North Africa and the Middle East were reputed to have been created by basilisks, whose glare was so terrible that all vegetation withered under it, and even solid rocks were split and sundered into sand.

Later reports of basilisks came in from Europe. As the centuries passed, the basilisk's form changed. Sometimes it was portrayed as a lizard with a rooster's head, or as a large lizard with six legs and a crown upon its head. The commonest form that these later basilisks took, was that of a huge rooster with the tail of a serpent or a lizard. Sometimes these beasts sported horns or antlers. In this form they were known as the cockatrice.

Its death-dealing powers came from its withering glare. Any creature that caught the eyes of the basilisk would fall dead from the uncanny power of its vision. There was but one exception to this, one animal that could withstand this 'look of death'. That animal was the weasel! It was believed that God never created a bane without creating some cure for it (like the stinging-nettle and the dock-leaf). Ergo, even the basilisk could be tackled by someone who knew its weaknesses. The monster's own gaze was as lethal to itself as to any other creature. Hence, its own reflection would kill it stone-dead! Equally—for some cryptic-reason—the sound of a cock crowing at dawn would also kill the basilisk.

In Pliny the Elders *Natural History,* written around 79 AD, the author places the creature in Libya.

> *There is the same power also in the serpent called the basilisk. It is produced in the province of Cyrene, being not more than twelve fingers in length. It has a white spot on the head, strongly resembling a sort of a diadem. When it hisses, all the other serpents fly from it: and it does not advance its body, like the others, by a succession of folds, but moves along upright and erect upon the middle. It destroys all shrubs, not only by its contact, but those even that it has breathed upon; it burns up all the grass, too, and breaks the stones, so tremendous is its noxious influence. It was formerly a general belief that if a man on horseback killed one of these animals with a spear, the poison would run up the weapon and kill, not only the rider, but the horse, as well. To this dreadful monster the effluvium of the weasel is fatal, a thing that has been tried with success, for kings have often desired to see its body when killed; so true is it that it has pleased nature that there should be nothing without its antidote. The animal is thrown*

into the hole of the basilisk, which is easily known from the soil around it being infected. The weasel destroys the basilisk by its odour but dies itself in this struggle of nature against its own self.

Saint Isidore of Seville covers the basilisk in his lengthy works.

'Basilisk' (basiliscus) is a Greek word, translated into Latin as 'little king' (regulus; king), because it is the king of the snakes, so that they flee when they see it because it kills them with its odour—it also kills a human if it looks at one. Indeed, no flying bird may pass unharmed by the basilisk's face but, however distant it may be, it is burnt up and devoured by this animal's mouth. However, the basilisk may be overcome by weasels. For this reason, people take weasels into caves where the basilisk lies hidden; and as the basilisk takes flight at the sight, the weasel chases it down and kills it. Thus, the Creator of nature sets forth nothing without a remedy. It is half a foot in length, and marked with white spots. Basilisks, like scorpions, seek after parched places, and when they come to water, they become hydrophobic and frantic. The sibilus (lit. 'the hissing one') is the same as the basilisk, and it kills by means of a hissing, before it bites and burns.

The Venerable Bede (672–735) was an English Benedictine Monk, traveller, and writer. He was the first to record the origins of the basilisk. He wrote that elderly roosters could lay eggs like hens. If one such egg was incubated by a snake or toad, a basilisk would hatch for it.

Alexander Neckam (1157–1217) was the abbot of Cirencester Abbey as well as a scholar, teacher, and theologian. He discounted the idea of the basilisk killing by its deadly gaze. Instead, he thought the animal gave forth an 'air of corruption' (venomous breath).

In the Middle Ages, alchemists believed that a mixture of powdered basilisk's blood, powdered human blood, red copper, and vinegar could transform copper into gold.

The Renaissance polymath Leonardo da Vinci includes basilisks in his bestiary. His thoughts are clearly influenced by Pliny.

This is found in the province of Cyrenaica and is not more than twelve fingers long. It has on its head a white spot after the fashion of a diadem. It scares all serpents with its whistling. It resembles a snake but does not move by wriggling but from the centre forwards to the right. It is said that one of these, being killed with a spear by one who was on horseback, and its venom flowing on the spear, not only the man but the horse also died. It spoils the wheat and not only that which it touches, but where it breathes the grass dries and the stones are split.

He also noted that the weasel kills the basilisk with the smell of its urine.

In Cantabria, on the north coast of Spain, the creature is called *basiliscu*. It was said to hatch from a leathery egg laid by an old rooster on the night of its death beneath a full moon. The resulting monster had a lizard's body and the head of a rooster. It could only be killed by hearing the sound of a cock crowing, ergo travellers used to carry live roosters with them in the area.

In 1202, one was found in a well in Vienna. Many people were killed by its venomous breath until it was finally dragged out. Another turned up in 1212 in the well in the courtyard of Schonlaterngasse 7, a house in a winding alleyway in central Vienna. The apprentice of a baker killed it by holding up a mirror up to it.

The basilisk was thought to spread plague via its breath and could set a whole community into a panic. One turned up in Rome during the reign of Pope Leo X (Pope from 1513–1521) and was blamed for an outbreak of plague.

In 1598, a basilisk made its lair in the cellar of an abandoned house in Warsaw and killed two young girls. It was slain by a man who entered the cellar clothed in a suit covered in mirrors. The basilisk's own evil gaze was as lethal to itself as to any other creature.

In 1640, a basilisk turned up in a well outside the Wight Hart Inn. Its breath killed several people before it was slain. The body is said to have resembled that of a snake.

These accounts could be caused by sulphurous fumes or methane gas rising up from underground and any snake in the area being transformed into the king of serpents by superstitious locals.

The basilisk may have had its genesis in a very real group of snakes: the cobras. As you may recall, the basilisk was vulnerable to only two creatures, the rooster and the weasel. Weasels belong to a group of mammals called mustelids. These include stoats, otters, badgers, wolverines, and martens. The group bears a striking similarity to the viverrids or mongooses. This is a case of convergent or parallel evolution, where two distinct groups of animals develop to resemble each other due to fulfilling similar ecological niches. There are differences. Mustelids have greater variation in form, and grow larger than viverrids, but the weasel and the common mongoose are sufficiently alike for confusion to arise between the two. The cobra's hood could also add to the legend by being distorted into the basilisk's crown. Mongooses are famed for killing venomous snakes.

Another factor in the legend of the basilisk seems to lie with the black mamba (*Dendroaspis polyepis*)—a large and highly venomous snake of sub-Saharan Africa. This snake can reach fourteen feet in length, and is renowned for its aggressiveness. When in a warning display, it can rear up to the height of a man, and unlike most other snakes it will actively pursue and attack anything that it perceives as a threat. Most snakes strike only once. The black mamba strikes repeatedly, and has a potent neurotoxic (nerve-paralysing) venom. It has been recorded on many occasions, that black mambas shedding their skin will sometimes retain a flap of old dead skin upon their head. This strongly resembles the crest of a cockerel, and may well have led to legends of crested serpents—both in Africa and elsewhere—via visiting foreigners.

But what of the other factors in the basilisk legend? It seems that these also have explanations within the realm of the natural rather than the supernatural. The miraculous egg-laying cockerel is not so fantastic as it at first sounds. There is a disease in fowl that causes a hen's ovaries to become infected. This prevents the production of the female hormone estrogen. Estrogen controls feminine characteristics, and when these are prevented from developing, masculine traits appear. These include developing a comb and wattle, crowing, and attempting to mount hens. If the victim recovers, it returns to its former feminine self and may lay once more; ergo a cock that lays eggs.

How about the snakes that sometimes slithered out of hens' eggs, to the mortification and horror of medieval cooks? The explanation is almost as grim as the original legend! Chickens often suffer from round worms (*Ascaris*); endoparasitic, internal creatures that are mainly passed out in the bird's droppings, but they can on occasion enter the reproductive-tract and be incorporated into an egg. In times past—when there were no stringent hygiene laws—this would have occurred far more often than today. Round worms can measure up to sixteen inches and one could readily imagine the terror evoked by cracking open an egg to find a writhing 'basilisk' within!

The basilisk has not entirely been banished into limbo; it had one last trick up its scaly sleeve. When the Spanish conquistadors first began to explore South America, they discovered a large lizard. It was bright green and bore a rooster-like crest on its head. They naturally called it a basilisk. *Basiliscus basiliscus* lacks its legendary counterpart's baleful stare, but it has a power almost as incredible. When alarmed, this two-foot lizard rears onto its hind legs and runs across the surface of rivers. Its elongate toes splay out, spreading its body weight. As long as it runs quickly, it does not break the surface tension of the water. Hence it is sometimes known as the Jesus Christ lizard.

An unknown species of African snake bears an uncanny resemblance to the basilisk. The crested crowing cobra is reported from central and southern Africa. This reptile is said to be twenty feet long, and grey or brown in colour. It has a scarlet crest like a rooster's comb on its head, as well as a pair of red wattles. Its cockerel-like attributes do not end there. The creature makes a noise very much like a cockerel's crowing, hence its name. It is said to be arboreal and highly venomous. Hyraxes seem to be its favoured prey. It also attacks humans, by lunging from overhanging-branches, and biting their faces. Some natives, when walking through forested areas, are said to carry pots of boiling-water on their heads to scald the attacking creature.

Doctor J. O. Shircore obtained some remains of a crested crowing cobra in 1944. A witch doctor in Malawi gave him a plate of bone from the crest with skin still attached, and several vertebrae and neck bones from at least two specimens. He described the plate thus:

> Its skeleton consists of a thin lanceolate plate of bone (one and a half inches by half an inch wide, at its broadest part) with a markedly rounded smooth ridge one-eighth of an inch wide, slightly overhanging both sides of the upper border, with a distinct voluted curve to the left. The lower border is sharp-edged and faintly ridged. The lateral surfaces are concave, throughout the long diameter. The whole fragment is eminently constructed for the insertion and attachment of muscles, much the same as the structure of the breast-bone of a bird. Some skin, part of which spreads smoothly above the plate, on one side, is red in colour: and attached to the lower angle is a dark wrinkled bit, which appears to be a remnant of head skin, all of which should be valuable for purposes of identification. A small portion of bone, tapering towards both ends, a half-inch long by one-eighth of an inch wide, is missing from the lower anterior border, including the tip—it was broken off for use in medicine by the witch doctor, from whom the specimens were obtained.

Shircore, it should be noted, was a medical doctor. His note appeared in the magazine *African Affairs*. The remains have never been identified and to my knowledge nobody knows of their current whereabouts.

In a 1962 letter to the publication *African Wild Life*, John Knott recounted his brush with an individual of what may be the same species. Whilst driving home from Binga in the Kariba area of Zimbabwe, (then Southern Rhodesia), in May 1959, he ran over a six-foot-long black snake. Upon investigation he discovered that the reptile had a symmetrical-crest on its head. The crest could be erected by way of five internal prop-like structures.

The crested crowning cobra seems to have a smaller counterpart in the Caribbean. The eminent Victorian naturalist Phillip H Gosse records it in *The Romance of Natural History, Second Series*. In 1845–1846 Gosse visited Jamaica where he first heard of the creature from a respected medical man.

> *They had seen, in 1829, a serpent about four feet in length, but of unwonted thickness, dull ochre in colour with well-defined dark spots, having on its head a sort of pyramidal helmet, somewhat lobed at the summit, of a pale red hue. The animal, however, was dead, and decomposition was already setting in. He informed me that the [natives] of the district were well acquainted with it; and that they represented it as making a noise, not unlike the crowing of a cock, and being addicted to preying on poultry.*

> *My friend, a Spanish informant, had seen the serpent with mandibles like a bird, with a cocks crest, with scarlet lobes or wattles; and he described its habits, perhaps from common fame rather than personal observation, as a frequenter of hen-roosts, into which it would thrust its head, and deceive the young chickens by its imitative physiognomy, and its attempts to crow.*

Jamaican resident Jasper Cargill offered a sovereign for any specimen of the snake but was not successful in obtaining one. Cargill himself had seen the elusive snake some years before as Gosse records.

> *When visiting Skibo, in St George's, an estate of his father's, in descending the mountain-road, his attention was drawn to a snake of dark hue, that erected itself from some fragments of limestone rock that lay about. It was about four feet long and unusually thick-bodied. His surprise was greatly increased on perceiving that it was crested, and that from the far side of its cheeks depended some red coloured flaps, like gills or wattles. After gazing at him intently for some time, with its head well erect, it drew itself in, and disappeared amongst the fragmentary rocks.*

Cargill's son shot a specimen some years later.

> *Some youngsters of the town came running to tell me of a curious snake, unlike any snake they had ever seen before, which young Cargill had shot, when out for a day's sport in the woodlands of a*

neighbouring penn. They described it as a serpent in all respects, but with a very curiously shaped head, with wattles on each side of its jaws. After taking it in hand and looking at it, they placed it in a hollow tree, intending to return for it when they should be coming home, but they had strolled from the place so far that it was inconvenient to retrace their steps when wearied with rambling.

When the youths returned the next day, the corpse was missing—presumably taken by some scavenger. When the tale was recounted to Richard Hill, his godson Ulick Ramsay, told him that he, too, had seen such a snake shortly before:

...not long previously, he had seen in the hand of the barrack-master-sergeant at the barracks of a Spanish town, a curious snake, which he, too, had shot amongst the rocks of a little line of eminences near the railway, about two miles out, called Craigallechie. It was a serpent with a curiously shaped head, and projections on each side, which he likened to the fins of an eel, but said were close up to the jaws.

The basilisk may have living, unknown analogues at large in the world today.

The Salamander

Like the basilisk, this creature possessed death-dealing powers out of proportion to its size. It was no more than a foot in length and shaped like a lizard. Its body was covered in star-shaped markings. The salamander could live in naked flame without the slightest harm to itself. It was also highly poisonous, spitting a foul foam from its mouth. This caused its victim's hair to fall out, and skin to wither, before death. The only animal immune to its venom was the pig. Pigs could eat salamanders with immunity, but if humans were to then eat the flesh of the pig, they would die, due to the venom accumulated in the swine's fat.

Pliny the Elder notes...

A salamander is so cold that it puts out fire on contact. It vomits from its mouth a milky liquid; if this liquid touches any part of the human body it causes all the hair to fall off, and the skin to change colour and break out in a rash.

The power of the salamander's toxin was truly immense. Should a salamander enter a pool, the water therein would be poisoned indefinitely. Alexander the Great (356–323 BC) was said to have lost two thousand horses and four thousand soldiers, when they drank from a stream that a salamander had crawled through. It was also believed that if a salamander came into contact with wood used for a baker's fire, then the bread would be contaminated.

Another strange quirk of the salamander was that it was supposed to be able to spin itself a cocoon out of a fireproof wool-like substance. This became known as salamander's wool. This fireproof 'wool' was much sought after. Pope Alexander III (appointed 1159–1181) was said to possess his own tunic of salamander's wool. The Byzantine Emperor, Manuel Comnenus (1120–1180), was said to have received a letter from the semi-mythical Prester John, a Priest-King who ruled over a mysterious land that some now believe to be Ethiopia. In the letter, Prester John speaks of salamanders, and how their wool is gathered and spun into cloth. When in need of a cleaning, garments made from the wool were cast into flames. Some think that salamander's wool was in fact asbestos.

The year was 1505. The five-year-old boy, Benvenuto Cellini (1500–1571), who would later become the famous Renaissance silversmith and artist, was drying clothes beside the fire in his home with his father, Giovanni. Suddenly, the pair saw something moving in the flames. A tiny creature was disporting itself in the midst of the hottest part of the fire. It resembled a small lizard with queer star-like shapes along its sides. The flames had no effect on the animal at all, and it scurried about on the way that a normal lizard would run up and down a wall.

Benvenuto's father called over to his sister and pointed out the creature to her. Then he smacked his son's ear, causing the boy to cry, and said:

My dear little son, I did not give you that blow on account of anything you have done wrong, but only that you may remember

that the lizard you saw in the fire is a salamander, a creature
that has never been seen by anyone else of whom we have
reliable information.

The salamander has been accepted by mainstream science and its remarkable powers explained. Salamanders exist. They belong to a group of tailed amphibians, (not lizards, which are reptiles), of the order Caudata. They are distributed worldwide (with the exception of sub-Saharan Africa and Antarctica). Some are mildly poisonous and display the fact with bright colours to deter predators. But none of them have the lethal venom attributed to them in medieval times. Salamanders often lurk inside damp logs and this may account for their alleged fire-walking powers. If such a log were tossed upon a fire, the little creatures would emerge and try to crawl to safety. As a damp log would not burn quickly most salamanders probably managed to escape the flames. As for the salamander's wool—it sounds very much like asbestos. The symptoms of salamander poisoning—withering of skin and loss of hair—sound very like asbestos poisoning. How this was first linked with the salamander is still unclear.

The Mongolian Deathworm

The Mongolian deathworm has haunted the legends of the Gobi for hundreds of years. It is said to lurk beneath the ancient sands of the Gobi Desert in Mongolia and is known to the desert nomads as *allghoi khorkhoi*. This translates as 'intestine worm', due to its resemblance to a length of cow's intestine. The creature is described as ninety to one hundred and fifty centimetres (three to five feet) long, blood-red in colour, and as thick around as a man's arm. It is apparently very hard to tell the animal's head apart from its tail. In the west, this horror has been given the name 'Mongolian deathworm' and with good reason. Mongolians go in great fear of the monster on account of its death-dealing capabilities. The deathworm can spit a corrosive venom that acts like acid, searing through the victim's flesh and into the veins. Those killed by the deathworm's venom are said to turn yellow—the allghoi khorkhoi's grotesque calling card. As if this were not enough, the deathworm is believed to generate a deadly electrical charge that can strike down victims from several metres away.

The deathworm first came to the attention of the west by the 1926 book *On the Trail of Ancient Man* by the legendary palaeontologist (and the real life inspiration for Indiana Jones) Prof. Roy Chapman Andrews. Andrews had led the 1922 Central Asiatic Expedition from the American Museum of Natural History. His main aim was to find fossil evidence of ancient man, but instead he found some of the most important dinosaur fossils ever discovered. These included the first dinosaur nest sites—complete with eggs—and some of the earliest fossil birds.

Before he embarked on this epic journey, Andrews met the Mongolian cabinet in order to obtain the necessary permits, Mongolia having previously been closed to outsiders. He was amazed to find no less a personage than the Mongolian Premier in attendance. The Premier had an unusual request for Andrews—to capture a deathworm should he come across one. He was even given forceps and dark glasses to protect him, (presumably from it spitting venom). Andrews commented on the death worm:

> *To the Mongols, it seems to be what the dragon is to the Chinese.*

Perhaps he was closer to the truth than he realised. He seemed, however, to doubt its existence.

> *This is probably an entirely mythical animal, but it may have some little basis in fact, for every northern Mongol firmly believes in it and gives essentially the same description. It is said to be about two feet long, the body shaped like a sausage, and to have no head or legs; it is so poisonous that even to touch it means instant death. It is reported to live in the most arid, sandy regions of the western Gobi. What reptile could have furnished the basis for the description is a mystery!*

The monster is mentioned in Ivan Antonovich Efremov's 1958 book *The Wind's Path*. (Efremov was a palaeontologist on a Soviet expedition to Mongolia in 1946.) Danzang—a geologist fluent in Mongolian—enquired about the deathworm, to an old man called Tseveng. The Mongol told him that its lair was a desolate wasteland called Khald-zan-dzakh, eighty miles to the southeast of Dalandzadgad. It lurked beneath the sands, surfacing only in June and July. Upon hearing of the worm's deadly powers, Danzang thought the whole tale was a joke. The old man replied in anger:

You laugh only because you know nothing and understand nothing. The allghoi khorkhoi—it is a terrible thing!

The men were impressed by the real fear displayed by Mongols in relation to the deathworm. Efremov himself believed the animal to be some kind of living fossil perpetuating its line from prehistory into historic times.

Czech explorer Jaroslav Mares heard an account of just how lethal the worm can be in 1967, whilst searching for dinosaur bones in Nemeght.

My brother—living in Oboto Chajun aimak—knew a man who encountered an allghoi khorkhoi', one herdsman told me. His name was Atlan. Once he returned with a friend from a neighbouring camp. They were riding their horses, and it was just after noon, one day in July. The sun was shining…

Suddenly Altan's friend's horse fell down. The rider stood up and went to the horse, but suddenly cried out and fell again. Atlan was five metres behind, and saw a big, fat worm slowly crawling away. Atlan stood in horror and then ran to his friend. But he was dead and so, too, was his horse.

The first westerner to have embarked on organised expeditions to specifically look for the deathworm was the Czech cryptozoologist Ivan Mackerle, who sadly passed away several years ago. Mackerle made two treks into the hostile sands of the Gobi hunting his deadly quarry.

The first took place in June and July of 1990 and concentrated on the desert southwest of Dlandzadgad. This area is so remote that not even Mongolian explorers have visited it. They used a low-tech method of attempting to attract the worms by driving a wooden log repeatedly into the ground. The worms were not impressed and did not appear.

Some interesting anecdotal evidence turned up, however. The expedition's interpreter, Sugi, told them of a dramatic incident from his childhood. A party of geologists had been visiting Sugi's home region. One of them was poking into the sand with an iron rod when he suddenly collapsed as if pole-axed. His colleagues rushed to his aid only to find him dead. As they examined the ground into which he had shoved, they saw the sand begin to churn violently. Out of the dune came a huge bloated deathworm.

An even more spectacular demonstration of the electrical potency of the deathworm came from no less an authority than a nature-ranger. The man, Yanzhingin Malhgalzahav (who hailed from Dalanszadgad), told the team that in the 1960s—just north of Noylon—a single deathworm had electrocuted a whole herd of camels, when one of the luckless creatures had stepped on it.

The second outing occurred in 1992 and was based to the west of Dalanszadgad. They interviewed lamas, shamans, and nomads close to the Chinese border. They also tried to force the worms to surface using controlled detonations—a little like fishing with dynamite but less wet. Once more the worms remained elusive. Perhaps the shock waves just frightened them deeper beneath the sands. The trip was filmed for Czech television as a documentary entitled *The Sand Monster Mystery*.

Despite the stars of the show failing to put in an appearance, more information was gathered on the beast. One old woman named Puret claimed that when on the attack, the worm rears up half its body like a cobra. A bubble of venom is formed at one end as the worm inflates itself. The bubble ultimately squirts forth at its victim. Anything struck by this noxious emission turns yellow and is corroded.

The question of the deathworm's identity is thorny. Some of the candidates of the former—true worms (annelids)—require moist conditions in which to live. They quickly become desiccated in hot, dry climates. It is far more probable that the creature is a vertebrate. A huge, limbless skink would make an excellent deathworm. Many of the desert-dwelling species have much reduced limbs and effectively swim through the sand like fish.

Another group of reptiles that match the worm's description even more closely are the amphisbaenas. Though they somewhat resemble primitive snakes or legless lizards, they are in fact neither. Amphisbaenas form a sub-order of squamata—the group of reptiles that includes snakes and lizards. Their exact relationship to the other sub-orders is far from clear.

As the sharp-witted reader will have noticed, these queer creatures derive their name from the amphisbaena of legend—the snake with a head at each end. Indeed, at first glance, it is hard to tell these animal's heads from their tails. They dwell in substrate, awaiting prey (mainly invertebrates) that they detect through vibrations, being virtually blind. The head is blunt and bullet-shaped, for pushing its way through the earth. The tail is almost identical in shape. When threatened, the amphisbaena will wave its tail complete with this 'false head' aloft. This diverts the attacker's attention from its true head. In some species the tail can be shed allowing its owner to make an escape. Many are pink or reddish and match the deathworm's description.

In colour and shape, amphisbaenas are dead ringers for our beast. The raising of the tail to enemies may explain the way the worm is said to rear up when spouting venom. But what of the venom? No known species of amphisbaenas is poisonous, but it is feasible that one could have evolved. Alternatively, it has been suggested that the 'venom' is in fact excrement of the creature's tail-end, rather than its head being raised. This sounds odd, but several species of animal defecate upon attackers. These include the grass snake (*Natrix natrix*) that squirts a yellowish-white, foul-smelling excreta onto those who harass it.

The largest known amphisbaenas grow to a length of around two feet—far shorter than some of the death worm's reported lengths. The worm's reported size and bulk would mean that if it was an amphisbaena it would be by far the largest worm lizard in the world.

Alternatively, the Gobi's most dangerous resident could be a snake. One group of snakes, known as sand boas (*Eryx spp*), fit the bill nicely as far as shape is concerned but are totally harmless. Some unknown venomous species would surely make a better model. Some snakes, such as the spitting cobras or rinkhals, can spray their venom over two metres (seven feet). The fangs have smaller apertures than those of non-spitting cobras. The venom is forced through them like water out of a hypodermic needle. Spitting cobras aim at the eyes of their enemies, and the venom can cause blindness. All cobras have the elongated body-plan of an average snake, however, unlike the squat, sausage-shaped deathworm. Perhaps the brute is more closely related to the Australian death adder (*Acanthophis antarcticus*). This snake is not a true adder (Viperidae) as none of these snakes inhabit Australia. Rather, it is a relative of the cobras, coral snakes, and kraits (Elapidae). With no true adders or vipers in the picture, a squat viper-like Elaphid snake evolved. This is known as convergent or parallel evolution, where two unrelated species in different parts of the world evolve to resemble each other because they inhabit similar ecological niches. Perhaps the death worm is a spitting Elaphid that has become short and squat in shape, much like the death adder.

The most problematic aspect of the deathworm is its alleged electrical generating abilities. The only known animals that can generate electricity at voltages high enough to kill humans are fish. Several species employ electricity mainly to find and kill prey. The most infamous of these is the electric eel (*Electrophorus electricus*) of tropical South America. This is not a true eel but a colossal knife fish (Gymnotidae). The electric eel can discharge a shock of up to six hundred and fifty volts—enough to kill a horse! The charge is generated via highly modified cells called electroplaques that are flattened and arranged in columns. There are up to ten thousand of these cells in an electric eel. If the death worm has such a power, it would be truly unique, as this apparatus is known only in aquatic creatures. Water is an excellent conductor of electricity, whereas sand is a poor one. It is possible that the 'worm' generates electricity via friction as it slithers through the sands and can transmit this over short distances like a living spark plug. More likely, the electrical powers are nothing more than an example of the mythification process.

The Tatzelwurm

In the alpine mountains of Austria, Bavaria, and Switzerland, a strange reptilian animal is occasionally reported. It is described as a cylindrical, scaly animal with a blunt head and powerful jaws. Its legs are greatly reduced, and some say it sports only a front pair of limbs.

It grows to some three feet in length and is greatly feared on account of its aggressive nature. It is believed to have a bite so venomous that it can kill a cow and can even breathe out poisonous gas. It is known variously as the Tatzelwurm (worm with feet), the Springwurm (jumping worm), and the Stollerwurm (tunnel worm). For clarity, I shall stick with the former name.

In 1723, naturalist Johann Jakob Scheuchzer published his book *Itinera per Hevetiae*. In it he records the killing of a strange monster by one Jean Tinner, who came upon the creature on Frumensburg mountain in Switzerland. The animal had a coiled, snake-like body six feet long, with a black and grey colouration. It reared up its neck, revealing a cat-like head. Tinner shot and wounded it with his musket. Together with his father, he managed to kill the monster, but sadly its remains were not retained.

Local farmers believed that the odd animal had been suckling milk from their cows. This seems like an odd habit for a reptile, but it is a theme that is often repeated in folklore.

The creature's existence was accepted as fact in the Alps, and it appeared in several books on alpine natural history and hunting alongside more familiar animals. Swiss naturalist Friedrich von Tschundi was convinced of the reality of the creature, and wrote in 1861:

> In the Bernese Oberland and the Jura, the belief is widespread that there exists a sort of 'cave worm' which is thick, thirty to ninety centimetres long, and has two short legs; it appears at the approach of storms after a long dry spell...
>
> In 1828, a peasant in the Solothurn canton found one in a dried-up marsh, and put it aside intending to take it to Professor Hugi. But in the meantime, the crows ate half of it.
>
> The skeleton was taken to the town of Solothurn, where they could not decide what it was and sent it to Heidelburg where all trace of it was lost.

In 1903, the Austrian Privy Councillor, A. von Drasenovich, was told by a close friend of an attack by the Tatzelwurm. The man, a professional hunter, was at an altitude of four thousand nine hundred feet near Murau in Steiermark, when he encountered the beast. It resembled a grotesque worm nineteen inches long by three inches thick, with four stubby legs. As he approached, the monster leapt at his face. He slashed at it with his knife in self-defence, but the blade could not penetrate its thick scales. The brute made six of these spectacular leaps at the man, before retreating into a crack in some rocks and disappearing.

A poacher and herdsman were hunting on Hochfilzenalm mountain in the south of Austria, when he saw a Tatzelwurm three feet long, as thick as a man's arm, and with two short legs, basking on some rocks. The hunter raised his rifle to shoot it, but the animal made a huge, arching leap at him. At this point he took to his heels.

Three years later, two travellers in the Mur valley came upon a singular carcass. It seemed to be a partial skeleton of a huge lizard around four feet long. The pair had the presence of mind to show the corpse to a local veterinary student. The student identified it as the remains of a roe deer (*Capreolus capreolus*). The finders were not convinced by his conclusions, but once again the bones went missing.

Two years later, in exactly the same spot, a young shepherd saw a giant lizard that so scared him he refused to work there for the whole summer.

In the 1930s, Dr Gerhard Venzmer and Hans Fulcher collected the evidence of sixty witnesses. All agreed that the Tatzelwurm was one to two feet long, cylindrical in shape, with the tail ending abruptly. It had a large blunt head that grew directly into the body with no narrowing in the neck area. The eyes were large, and the body scaled. It hissed like a snake.

There have been sightings farther south in Europe as well. Similar creatures have been reported from France. In Ossum, a woman encountered one whilst picking berries in 1939. The late Roger Hutchings—a much-missed member of the CFZ—heard talk of such reptiles whilst living in Provence in the mid-'60s. The older people referred to the creature as 'Arasas' others called it 'Le Gros' (the big one). Most people claim to have seen it. Those with a rural existence such as shepherds and truffle gatherers were very familiar with it. They reported that it was seen basking close to holes or clefts in rocks. If disturbed, it would quickly retreat into its lair. The descriptions fit exactly with those given by witnesses further north.

Farmers in Palermo, Sicily, reported a snake-like creature with a cat-like head and two short legs, attacking pigs in 1954. The cat's head seems at odds with other reports, but perhaps this was in reference to its large eyes. Maybe the Siracuse animal and Jean Tiller's cat-headed horror mentioned earlier were of the same species.

Strangely, there is a report of what sounds very much like a Tatzelwurm from Denmark in June 1973! A strange 'snake' was reported from the Ulushale forest on Mon Island. It was described as four and a half feet long, with a dark back, light underside and a 'nasty' head. A description fitting our monster.

So, if it exists, what is the Tatzelwurm? There are several candidates. The ones that spring to mind most readily are the skinks. These are a group of elongated lizards. Most are tubular in shape and some resemble animated salamis! Many have greatly reduced or even vestigial limbs. One type—*Chalcides striatus*—dwells in the French maritime Alps and can reach two feet in length. This species is, however, quite slender in build.

The largest species is the monkey tailed skink (*Corucia zebrata*) of the Soloman Islands. It can grow to around two and a half feet, and with is heavy scales and blunt head it resembles the Tatzelwurm in passing. Could an even larger skink with much reduced limbs exist in the Alps?

Another possibility is that the Tatzelwurm is related to the European legless lizard (*Opisaurus apodus*), a large powerful legless lizard closely related to the well-known slow worm (*Anguis fragilis*). It can grow to almost four feet long and can inflict a mighty (but not venomous) bite. Most become hand tame in captivity, but a wild specimen will—if annoyed—hiss, rear, thrash, and savagely bite. This aggressive behaviour is reminiscent of the Tatzelwurm. The legless lizard's range is southeastern Europe and southwestern Asia. Perhaps our beast could be a stouter northern European relative of the legless lizard.

The deadly venom attributed to the Tatzelwurm may very well be folkloric—much like the capabilities attributed to the salamander. But there are genuine venomous lizards. The Mexican beaded lizard (*Heloderma horridum*), and the gila monster (*Heloderma suspectum*) both have a poisonous bite. The venom is painful, but not usually fatal to humans.

The alpine horror may indeed be a giant lizard with a toxic bite, probably the continent's largest reptile.

The Tatzelwurm may be something other than a reptile. An Austrian schoolmaster who came across one in 1929 whilst exploring a cave on the Tempelmauer believed it to be a giant salamander.

> *I started to look for the entrance to the cave. Suddenly, I saw a snake-like animal sprawled on the rotting foliage that covered the ground. Its skin was almost white, not covered by scales but smooth. The head was flat and two very short feet on the fore part of the body were visible. It did not move but kept staring at me with its remarkably large eyes. I know every one of our animals at first glance and knew that I faced one that is unknown to science, the Tatzelwurm. Excited, joyful, but at the same time somewhat fearful, I tried to grab the animal, but I was too late. With the agility of a lizard, the animal disappeared into a hole and all my efforts to find it were in vain. I am certain that it was not my imagination that let me see the animal but that I observed with a clear head.*

> *'My' Tatzelwurm did not have large claws but short and atrophied looking feet; his length did not exceed forty or forty-five centimetres. Most probably the Tatzelwurm is a rare variety of salamander living in moist caves and only rarely coming to the light of day.*

There are aquatic amphibians with only two legs. The siren (*Siren lcertina*) of the southeast United States is one such creature. Others have a pallid hue, such as the blind cave-dwelling olm (*Proteus anguinus*) of southeastern Europe. Both of these animals sport feathery external gills, a feature notably lacking in the schoolmaster's description. The white skin and troglodyte existence do suggest that what this man saw was some unknown cave-dwelling salamander. The large eyes are somewhat out of character. Creatures evolved for a cave existence usually have atrophied or even absent eyes. Perhaps the milky hide was due to albinism.

The Beast of Gévaudan

Sometimes a cryptozoological case is so horrific and strange that is seems like some horror novel come to life. Thus it was with the case of the Beast of Gévaudan. In April of 1764 a girl herding cattle in eastern Gévaudan, a historical area of southeastern France, was attacked by a 'monster'. Luckily for her, the bulls charged the creature and drove it back into the forest from where it had come. On June 30 of the same year, a fourteen-year-old shepherdess call Jeanne Boulet was not so lucky. Whilst she tended her flock in the hills near Les Hubacs, Jeanne was killed, torn apart, and devoured by a predator of phenomenal power. Thus started a three-year nightmare for the Gévaudan. Between 1764 and 1767, the 'beast' as it became known, killed one hundred and thirteen men, women, and children. Many more were mauled and countless livestock eaten. The scenario played out like the script of a horror movie, but it was only too real and unlike a film it had no neat conclusion.

In the intervening two hundred and fifty-four years, investigators and writers have tried to figure out just what killed those people. Suggested identities for the Beast of Gévaudan have included an outsized wolf, a wolf-dog hybrid, a human serial killer, a werewolf, a hyena, and some sort of pre-historic beast surviving in the remote wilderness.

It is important to look at the historic situation at the time. France had just lost a costly conflict, the Seven Years' War and the Gévaudan had suffered a series of cripplingly harsh winters. Disease had ravished the livestock in the area. In the already beleaguered peasant communities, news of the beast spread as it claimed more and more victims. The corpses of its prey were hideously mauled, some decapitated and most eaten or partly eaten. The populace was used to man-eating wolves but to them, this was clearly something different, something more formidable. Gabriel Florent de Choiseul Beaupre, Bishop of Mende, declared that the beast was a scourge sent by God as a punishment for their sins.

With the body count rising, the depredations reached the ears of King Louis XV. The king sent two of his foremost huntsmen, Jean Charles Marc Antoine Vaumesle d'Enneval and his son Jean-Francois, to deal with the beast. The pair led soldiers including dragoons and mounted infantry as well as peasants who had been given temporary permission to carry weapons, in hunts, all to no avail. The monster evaded them and the killing continued unabated. Due to their failure, they were replaced the following year by Francis Antoine, the King's gun bearer. On September 20, 1765 Antoine shot a one-hundred-and-thirty-pound wolf that he declared was the beast. The carcass was brought back to court with much fanfare and put on display. Antoine was hailed as a national hero and revived money and titles, but this was no deliverance. He had not slain the beast. The survivors of the attacks who saw the body declared that it was not the creature that had attacked them. Soon after, the monster struck again.

The creature avoided organised hunts and shunned poisoned bait. The beast killed with impunity and its reputation was now spreading to the rest of world. On June 19, 1767, peasant hunter Jean Chastel claimed to have shot the beast. He had killed a large, russet-coloured wolf-dog hybrid. Once more, survivors and witnesses were adamant that the creature was not the beast that had attacked them.

By the end of 1767, the story peters out. There was no final confrontation with the monster. It just vanished, quietly slipping into legend.

Wolves very rarely attack humans in modern times. Before the twentieth century, there were more attacks, but these focused mainly on children. Analysing the beast's prey selection, the German mammologist Karl Hans-Taake found that it targeted adult humans far more than wolves did, in fact six times more. The beast seemed to be targeting larger-bodied prey to feed a predator bigger than a wolf. Adult victims had most of their flesh eaten, beyond the stomach capacity of a wolf.

Looking at the attack mode of the beast, it seems that a wolf was again an unlikely candidate. Sometimes the victim would be decapitated, something wolves rarely do. Victims were also killed by throttling. The beast would put pressure on the trachea with its jaws but without puncturing in. One youth who survived such an attack was rendered mentally disabled due to lack of oxygen to the brain. His chest and scalp were racked by the monster's claws. The beast employed this mode of killing with livestock too. It was reported to leap upon the backs of cows, drag them down and throttle them with its mouth. No wolf is capable of this. The claws of a wolf are blunt and not employed in killing. The beast, however, ripped open both humans and animals with razor-sharp claws.

The size of the animal was variously said to be anywhere from that of a calf to a cow, but all agreed it was significantly bigger than a wolf. Cows in rural France at the time were mainly the Limousin breed and were no more than three feet at the withers and six to seven hundred pounds. The beast was shot many times but never killed. On one occasion on October 9, 1764, it took four musket shots but rose again to hobble away and lick its wounds. An animal bigger than a wolf with a more robust muscle and bone structure would be harder to kill. Etienne Lafont, a local official who coordinated many hunts, was sure the beast was not a wolf and tried to identify it from books on animals.

The strength of the beast's bite was in evidence in the case of Catherine Valley, a sixty-year-old widow who had her head bitten in two like a cracked nut. The skull had been cleaned of flesh both inside and out, as if polished with a tool, another clue as to the nature of the beast.

On the 13th of March 1765, at the hamlet of La Bessiere, thirty-five-year-old mother of four, Jeanne Valet, was brave enough to leap on the back of the beast when it seized her six-year-old son. Apparently, the brave woman mounted it as if it were a horse and succeeded in making it let go of her son by grabbing at its testicles. She was badly clawed and scalped for her efforts. When her older son arrived, he attacked the beast with a long stick ending in a blade. His large dog tackled the monster as well, biting at its head before being thrown ten feet by the angry beast. The beast finally retreated. Jeanne Valet survived despite her wounds but her younger son died six days later. No wolf is big enough to be ridden like a horse (despite what you may have read in *The Hobbit*).

In another occurrence, the beast was attacked by six huge hounds trained in wolf hunting. The monster fought back with both teeth and huge claws, killing two hounds and escaping. On several occasions, the beast attacked people on horses, seemingly just as interested in killing the mount as the rider. Some of these horses suffered long slashes from the attacker's claws.

The last victim of the beast was nine-year-old Catherine Chautard on June 9, 1767, near Coffours. Author of *The Gévaudan Tragedy*, Karl Hans-Taake, thinks it may have succumbed to the extensive campaign of leaving out poisoned bait or possibly simply stopped preying on humans and subsisted for the rest of its days on the wild game that had been part of its diet the whole time.

The descriptions of the beast give an identikit. Most said it was as big as a yearling calf. The fur was russet or reddish-brown with a white patch on the broad chest. It had powerful front quarters and a broad, flat head with erect ears. The muzzle was broad like a calf's and had large teeth. The beast was armed with claws and had a long tail, thick as a human arm that ended in a bushy tassel. Its call was described as a moaning roar. Taken together with the mode of killing and feeding, there can be little doubt that the Beast of Gévaudan was a sub-adult male lion (*Panthera leo*).

One may ask that if the beast was a lion, then why was it not described as such? Most people in rural France in the eighteenth century would only be familiar with lions from illustrations in books which mostly depicted them with full manes. Male lions don't generally develop full manes until they are around four years old. The beast was sometimes described as having a tuft of hair on the head like a mohawk haircut. Sub-adult lions have these tufts. Some adult male lions lack manes altogether such as the infamous man-eating lions of Tsavo that terrorised workers on the Kenya-Uganda railway in 1898.

Where did the beast come from? Trade in wild animals for menageries had been going on for many years. Lions, tigers, bears, and even crocodiles had been displayed at the Tower of London. The Royal Menagerie began as far back as 1204. In France, the Royal Menagerie at Versailles was established in 1661. It held lions, leopards, and hyenas. Interestingly, there had been other cases of man-eating beasts in France. For example, from 1698 to 1700, another beast with a description that matches the more famous later case killed many men women and children and sprang at horses. This was in the Limousin region of France, close to Gévaudan. Another region neighbouring Gévaudan, the Dauphine, saw similar attacks from 1752 to 1756 and from 1762 to 1763, these included the beheading of victims. I have to note that these attacks rose with the popularity of the French aristocracy keeping private zoos on their lands. The creature that killed one hundred and thirteen people almost certainly escaped from a captive collection or escaped during transport to a collection. The Beast of Gévaudan was a man-made monster.

It is interesting to note how a known animal, taken from its natural habitat and placed in an alien one, becomes a 'monster'. This has happened many times. Big cats loose in the UK have been dubbed 'The Beast of Bodmin', the 'The Beast of Bevendean', 'The Beast of Exmoor', and many more dramatic names. I myself have seen one of these creatures on the outskirts of Exeter.

I was travelling by coach from Exeter to Bristol on the M5. It was late May or early June 2011. It would have been around noon. I looked out of the window to my left. In a field, close to a large tree, stood a massive cat. It was roughly the size of an Alsatian dog in terms of height but somewhat longer in the body. It was brown in colour and had a long tail that extended almost to the length of the head on the body. The head was round with rounded ears. The animal stood totally still and I had it in view for around ten seconds. It was about five hundred feet away from me at the time. No one else on the coach seemed to notice it. I am a former zoo keeper and am familiar with big cats. The animal appeared to be an adult puma (*Puma concolor*). Thankfully, these smaller cats have not placed humans on their menu.

Afterword

And so we have reached the end of volume one. Originally, I was planning a single volume, but the material was so vast that the publishers and I decided to split the tome into two parts. To be honest, even with this I have hardly scratched the surface of this subject. I could have written a series of books, each of encyclopaedic length on cryptozoology. I hope, however, that I have piqued the reader's interest.

The evidence for many of the creatures covered in this book is now mounting, but mainstream science still remains largely hostile to the subject. My friend, Adam Davies, once prepared a paper with Professor Hans Brunner, a world recognised expert in mammal hair. The subject was their findings after the analysis of suspected Orang-pendek hairs from Sumatra. The paper was submitted to the magazine *Nature*. The editor refused to publish the paper on the grounds that it was dealing with a large, unknown animal. He told them that if it was about a new species of mouse then there would have been no problem. What an unscientific and ridiculous approach to data! The attitude beggars belief.

It seems that if a theoretical physicist writes a paper on a molecule that nobody has ever seen, but its existence is inferred by the reactions of other molecules, then that is fine. But should a zoologist write a paper on a creature that has been seen by thousands of people but for which we do not yet have a type specimen, then he is a scientific heretic. These double standards can do nothing but harm science in the long run.

Remember the fate of the sorely wronged Pierre Denys de Montfort? Little has changed two hundred years since his death. It is up to people like you and I, dear reader, to change that. As Bernard Heuvelmans, once said: 'The great days of zoology are not done'. The discoveries are to be made in the jungles, mountains, deserts, and oceans—not in the lecture halls. There are still wild and unexplored places on earth and there are still unknown creatures inhabiting them. When your mother and father told you there were no such things as monsters, they were wrong.

In volume two, I will examine the possible survival of species thought to be extinct such as the Tasmanian wolf, giant ground sloths, and supposed living dinosaurs. I will look at specimens of know creatures that have grown far beyond the limits set upon them by textbooks. These include giant crocodiles and snakes. Also, I will be telling the reader of my own adventures hunting cryptids all across the globe, including the Tasmanian wolf, the Yeti, the Mongolian deathworm, the Almasty, the Orang-pendek, and more. Finally, I will be giving advice, in practical steps, to those of you who would like to organise your own cryptozoological expedition. Who knows...*you* may be the one to make the breakthrough.

RICHARD FREEMAN
EXETER, UK, JANUARY 2019

Acknowledgements

Many thanks to the following people. Shawn Hoult, Hugo Villabona, and all at Mango for coming up with the idea for this book. Sir David Attenborough for childhood inspiration. Lars Thomas for the foreword and all his tireless work for the Centre for Fortean Zoology. Jon Downes and the crew at the CFZ for flying the Fortean flag for all these years. Rob Morphy, Marc Storrs, and Chris Carnicelli at *The Cryptonaut Podcast* which I listen to whilst proof editing. Jackie Tonks for her account of a Sasquatch encounter. Jon Pertwee, Tom Baker, Darren McGavin, and Ray Harryhausen, without whom I would never have become a cryptozoologist. And finally, to all my fellow cryptozoologists around the world, the science Bohemians.

Bibliography

Bailey, Col, *Shadow of the Thylacine*, (Five Mike Press, 2013)

Bayanov, Dimitri, *In the Footsteps of the Russian Snowman* *(Cryptologos*, 2004)

Blackburn, Lyle, *The Beast of Boggy Creek* (Anomalist Books, 2013)

Blackburn, Lyle, Beyond Boggy Creek (Anomalist Books, 2017)

Blashford-Snell, Colonel John, *Mysteries* (Bodley Head,1983)

Bondeson, Jan, *The Feejee Mermaid and other Essays in Natural and Unnatural History* (Cornell University Press, 1999)

Bord, Janet and Colin, *Alien Animals* (Granada, 1980)

Bord, Janet and Colin, *The Bigfoot Casebook* (Granada, 1982)

Bord, Janet and Colin, *Modern Mysteries of the World* (Grafton Books, 1989)

Bord, Janet and Colin, *Modern Mysteries of Britain* (Grafton Books, 1988)

Bright, Michael, *There are Giants in the Sea* (Robson Books, 1991)

Byrn, Peter, *Hunt for the Yeti* (CreateSpace Independent Publishing Platform, 2015)

Coleman, Loren and Clark, Jerome, *Creatures of the Outer Edge* (Warner Books, 1978)

Coleman, Loren, *Curious Encounters* (Faber and Faber, 1985)

Coleman, Loren, *Mysterious America* (Faber and Faber, 1983)

Coleman, Loren, *Tom Slick and the search for the Yeti* (Faber and Faber, 1989)

Corliss, Richard, *Incredible Life: A Handbook of Biological Mysteries* (Sourcebook Project, 1981)

Costello, Peter, *In Search of Lake Monsters* (Garnstone Press, 1974)

Couzens, T, *Tramp royal: The true story of Trader Horn.* (Raven Press, 1993)

Debenat, Jean-Paul, *The Asian Wildman* (Hancock House, 2014)

Dickenson, Peter, *The Flight of Dragons* (Pierrot Publishing, 1979)

Dinsdale, Tim, *The Laviathans* (Routledge & Kegan Paul, 1966)

Dinsdale, Tim, *The Story of the Loch Ness Monster* (Target, 1973)

Eberhart, George M, Monsters: I*ncluding Bigfoot, Many Water Monsters, and Other Irreguar Animals* (Garland Publihing, 1983)

Eberhart, George M, *Mysterious Creatures: A Guide to Cryptozoology* (ABC-CLIO, 2002)

Freeman, Richard, *Dragons: More Than a Myth?* (CFZ Press, 2005)

Freeman, Richard, *Explore Dragons* (Heart of Albion, 2006)

Freeman, Richard, *Orang-Pendek: Sumatra's Forgotten* Ape, 2011)

Fourth, Gregory, *Images of the Wildman in Southeast Asia* (Routledge, 2012)

Furneaux, Robin, *The Amazon* (Reader's Union, 1971)

Gaal, Arlene, *Ogopogo, The* True Story of the Okanagan Lake Million Dollar Monster (Hancock House, 1986)

Gaal, Arlene, *In Search of Ogopogo: Sacred Creature of the Okanagan Waters* (Hancock House, 2001).

Gilroy, Rex, *Mysterious Australia* (Nexus Publishing, 1995)

Gould, Charles, *Mythical Monsters* (W H Allen & Co, 1886)

Gosse, Philip Henry, *The Romance of Natural History: Second series* (James Nesbit and Co, 1866)

Guggisberg, C.W. A, *Crocodiles: their Natural History, Folklore, and Conservation* (David & Charles, 1972)

Healy, Tony, and Cropper, Paul, *Out of the Shadows, The Mystery Animals of Australia*, (Ironbark, 1994)

Healy, Tony, and Cropper, Paul, *The Yowie: In Search of Australia's Bigfoot* (Anomalist books, 2006)

Harrison, Paul, *Sea Serpents and Lake Monsters of the British Isles* (Hale, 2001)

Harrison, Paul, *The Encyclopaedia of the Loch Ness Monster* (Hale, 1999)

Hegeenbeck, Carl, *Beasts and Men Longmans*, (Green and Co, 1909)

Heuvelmans, Dr Bernard, *On the Track of Unknown Animals* (Rupert Hart-Davis, 1958)

Heuvelmans, Dr Bernard, *In the Wake of the Sea Serpents* (Rupert Hart-Davis, 1968)

Holiday, FW, *The Great Orm of Loch Ness* (Faber & Faber, 1968)

Hughes, J. J. *Eighteen Years on Lake Bangweulu* (1933, The Field)

Izzard, Ralph, *Abominable Snowman Adventure* (Hodder and Stoughton, 1955)

Kearton, Cherry, *In the Land of the Lion* (National Travel Club, 1929)

Kingsley, Mary, *Travels in West Africa* (Mcmillan & Company, 1897)

Knight, Charles, *Pictorial Museum of Animated Nature* (London, 1844)

Kirk, John, *In the Domain of the Lake Monsters* (Key Porter Books, 1998)

Landsburg, Alan, *In Search of Myths and Monsters* (Corgi, 1977)

Lee, John and Moore, Barbara, *Monsters Among Us: Journey to the Unexplained* (Pyramid, 1975)

Marshall, W. H, *Four Years in Burma* (C. J. Skeet, London, 1860)

Mackal, Dr Roy, *Searching for Hidden Animals* (Cadogab Books, 1983)

Mackal, Dr Roy, *A Living Dinosaur? In Search of Mokele-Mbembe* (E. J. Brill, 1987)

McEwan, Graham J., *Mystery Animals of Great Britain and Ireland.* (Robert Hale, 1986)

Montgomery Campbell, Elizabeth, and Soloman, David, *The Search for Morag* (Tom Stacey, 1972)

Morris, Desmond and Morris, Ramona, *Men and Snakes* (Sphere, 1965)

Newman, Paul, *The Hill of the Dragon* (Kingsmead Press, 1979)

Sanderson, Ivan T, *Abominable Snowmen: A Legend Come to Life* (Chilton Company, 1961)

Sanderson, Ivan T, *Things* (Pyramid Books, 1968)

Sanderson, Ivan T, *More Things* (Pyramid Books, 1969)

Shackley, Myra, *Still Living? Yeti, Bigfoot and the Neanderthal Enigma* (Thames and Hudson, 1986)

Shuker, Dr Karl, *The Lost Ark* (Collins, 1993)

Shuker, Dr Karl, *The Unexplained* (Carlton, 1996)

Shuker, Dr Karl, *In Search of Prehistoric Survivors* (Casselle Illustrated, 1997)

Shuker, Dr Karl, *From Flying Toads to Snakes with Wings* (Llewellyn Publications, 1997)

Shuker, Dr Karl, *Meet Mongolia's Death Worm* (in Fortean Studies Volume 4, John Brown Publishing, 1998)

Shuker, Dr Karl, *Mysteries of Planet Earth* (Carlton,1999)

Shuker, Dr Karl, *Dragons in Zoology, Cryptozoology and Culture* (Coachwhip Publications, 2013)

Simpson, Jacqueline, *British Dragons* (BT Batsford, 1980)

Smith, Malcomb, *Bigfoots and Bunyips: In search of Australia's Mystery Animals* (Millenium Books, 1996)

Steel, Rodney, *Crocodiles* (Helm, 1989)

Sykes, Bryan, *Nature of the Beast* (Coronet, 2015)

Thomas, Lars, *Weird Waters: The Lake and Sea Monsters of Scandinavia and the Baltic States* (CFZ Press, 2011)

Thomas, Lars, *Curious Countries: The Mystery Animals of Scandinavia and the Baltic States* (CFZ Press, 2019)

Waterton, Charles, *Wanderings in South America* (B. Fellows, 1839)

Whitlock, Ralph, *Here be Dragons* (George Allen & Unwin, 1983)

Williams, Mike and Lang, Rebbeca, *The Tasmanian Tiger, Extinct or Extant?* (Strange Nation Publishing, 2014)

Willoughby-Meade, G, *Chinese Ghouls and Goblins* (Constable, 1928)

Xu, David C, *Mystery Animals of China* (Coachwhip Publications, 2018)

About the Author

Richard Freeman is a full-time cryptozoologist. He searches for and writes about unknown animals. He has hunted for creatures such as the Yeti (a dark haired, giant, upright ape in North India), the Mongolian deathworm (a much feared burrowing reptile of the Gobi), the giant anaconda (a monster constricting snake in South America), the Ninki Nanka (a dangerous dragon-like beast from the swamps of West Africa), the Almasty (a relic hominid in the Caucasus of Russia), Orang-pendek (an upright-walking ape in Indonesia), the Naga (a giant, crested serpent in Indo-China), the Gul (a relic hominin from Tajikistan) and the Tasmanian wolf (a flesh-eating marsupial in Tasmania). He is the zoological director at the Centre for Fortean Zoology. This is the world's only full-time mystery animal research organisation. It is based in North Devon.

He has lectured at the Natural History Museum in London and the Grant Museum of Zoology. Richard is also a regular contributor to the magazine *Fortean Times*.

He has written books about cryptozoology, folklore, and monsters including *Dragons: More Than a Myth?*; *Explore Dragons*; *The Great Yokai Encyclopaedia: An A to Z of Japanese Monsters*; and *Orang-Pendek: Sumatra's Forgotten Ape*. However, he has recently branched out into horror and weird fantasy with *Green Unpleasant Land: 18 Tales of British Horror* and *Hyakumonagatari: Tales of Japanese Horror, Book One*. This book is his latest work, an overview of cryptozoology and a chronicle of his own expeditions.

Richard is a massive fan of classic *Doctor Who* ('60s/'70s) and a lover of weird fiction and horror.

Mango Publishing, established in 2014, publishes an eclectic list of books by diverse authors—both new and established voices—on topics ranging from business, personal growth, women's empowerment, LGBTQ studies, health, and spirituality to history, popular culture, time management, decluttering, lifestyle, mental wellness, aging, and sustainable living. We were recently named 2019 *and* 2020's #1 fastest growing independent publisher by *Publishers Weekly*. Our success is driven by our main goal, which is to publish high quality books that will entertain readers as well as make a positive difference in their lives.

Our readers are our most important resource; we value your input, suggestions, and ideas. We'd love to hear from you—after all, we are publishing books for you!

Please stay in touch with us and follow us at:

Facebook: Mango Publishing
Twitter: @MangoPublishing
Instagram: @MangoPublishing
LinkedIn: Mango Publishing
Pinterest: Mango Publishing

Newsletter: mangopublishinggroup.com/newsletter

Join us on Mango's journey to reinvent publishing, one book at a time.

9 781642 500158

Printed in the USA
CPSIA information can be obtained
at www.ICGtesting.com
JSHW031133210923
48853JS00002B/3

9 781642 500158